DRIVING STANDARDS AGE

THE Goods Vehicle Driving Manual

DRIVING SKILLS

Published by The Stationery Office Limited

Fourth impression 1999

This title was formerly known as
Your Large Goods Vehicle Driving Test

First edition Crown copyright 1994
Second edition Crown copyright 1995

ISBN 0 11 551783 9

British Library Catalogue in Publication Data
A CIP catalogue record for this book is available from the
British Library

Other titles in the Driving Skills series

The Complete Theory Test for Car Drivers and Motorcyclists
The Official Theory Test for Drivers of Large Vehicles
The Driving Test
The Driving Manual
The Bus and Coach Driving Manual
The Motorcycling Manual
The Theory Test and Beyond (CD-ROM)

ACKNOWLEDGEMENTS

The Driving Standards Agency (DSA) would like to thank the staff of the following organisations for their contribution in the production of this publication

Driver & Vehicle Testing Agency Northern Ireland

Driver and Vehicle Licensing Agency

Department of Transport

Traffic Director for London

London Borough Cycle Officers

Association of Police Officers in Scotland

Road Haulage and Distribution Training Council

Road Haulage Association

The Driving Standards Agency (DSA) is an Executive Agency of the Department of Transport. You'll see its logo at test centres.

The aim of the DSA is to promote road safety through the advancement of driving standards.

DSA

- Conducts practical driving tests for drivers or riders of cars, motorcycles, lorries, buses and other vehicles
- Plans, maintains and supervises the theory test for drivers or riders of cars, motorcycles, lorries and buses
- Controls the register of Approved Driving Instructors (ADIs)
- Controls the voluntary register of Large Goods Vehicle (LGV) Instructors
- Supervises Compulsory Basic Training (CBT) courses for motorcyclists
- Aims to provide a high-quality service to its customers

CONTENTS

Acknowledgements iii
About the Driving Standards Agency v
Introduction 1
About this book 2

PART ONE **Getting Started**

- Getting started 6
- The LGV driver 12
- Attitude 14
- Driving forces 19
- Vehicle characteristics 28

PART TWO **Limits and Regulations**

- Vehicle limits 37
- Braking systems 46
- Load restraint 52
- Environmental impact 57
- Legal requirements 62
- Other regulations 70

PART THREE **Driving Skills**

- Professional driving 81
- Driving at night 93
- Motorway driving 100
- Bad weather 116
- Accidents 124
- Breakdowns 134

PART FOUR **Preparing for the Driving Test**

- About the driving test 136
- Theory into practice 139
- How to apply for your test 140
- Before attending your test 142
- Legal requirements 146
- At the test centre 147
- The official syllabus 148

PART FIVE The LGV Test

- The reversing exercise 161
- The braking exercise 163
- The vehicle controls 164
- Other controls 170
- The gear-changing exercise 171
- Moving off 172
- Using the mirrors 175
- Giving signals 176
- Acting on signs and signals 177
- Awareness and anticipation 178
- Making progress 179
- Controlling your speed 180
- Separation distance 181
- Hazards 182
- Selecting a safe place to stop 201
- Uncoupling and recoupling 202
- If you pass 204
- If you don't pass 205

PART SIX Additional Information

- Disqualified drivers 207
- DSA services 208
- DSA Area Offices 210
- LGV test centres 211
- Traffic Area Offices 212
- Other useful addresses 214
- Minimum test vehicles (MTVs) 215
- Hazard labels 216
- Glossary of terms 218

INDEX 223

As a large goods vehicle (LGV) driver you have a special responsibility – not just to yourself, but to all other road users. A professional driver should set an example to other drivers by ensuring that the vehicle is driven, at all times, with the utmost safety and with courtesy and consideration for everyone else on today's busy roads.

To become an LGV driver you must possess a high degree of skill in the handling of your vehicle and also be prepared to make allowances for the behaviour of others. The right attitude and approach to your driving, together with a sound knowledge of professional driving techniques, and the ability to apply those techniques, are essential.

By successfully passing your car driving test you've already shown that you've reached the standard set for driving a motor vehicle unsupervised on today's roads. This book sets out the skills that you must now show in order to pass the vocational driving test. Put the information it contains into practice and you should be able to reach the higher standards demanded. From there you could earn the privilege of driving large goods vehicles and, above all, go on to a career of '*Safe driving for life*'.

Robin Cummins
Chief Driving Examiner
Driving Standards Agency

This book will help you to

- **Understand what a large goods vehicle (LGV) driver needs to know to be thoroughly professional**
- **Prepare you for the range of skills you'll need to show to pass your practical LGV driving test**

Part One tells you how to get started. Here you'll also learn about the difference between driving lorries and cars. The importance of driver attitudes and different vehicle characteristics are also covered.

Part Two covers the various rules and regulations you'll need to know and observe. Vehicle limits, braking systems, load restraints, and legal requirements are all discussed. The effect of your vehicle on the environment as well as health, conduct and safety issues are also tackled.

Part Three tells the potential professional driver about driving in different weather conditions, at night and on motorways, and about dealing with accidents and emergencies. It also looks at the essential skills you'll need to drive professionally.

Part Four contains the official syllabus for learning to drive an LGV and shows the skills you need to learn before taking your test.

Part Five explains the test requirements and provides clear advice. Refer to it regularly and use it to check your progress.

Part Six gives additional useful information.

The important factors

This book is only **one** of the important factors in your training. Other factors are

- **A good instructor**
- **Plenty of practice**
- **Your attitude**

If you're driving an LGV you must ensure that your goods arrive at their destination safely. This process will not only involve the safety of your load, but also your attitude to others on the road. From the start, you must be aware of the differences between driving small and large vehicles.

To become a professional driver you must have a thorough knowledge of the regulations that apply to your work.

It's also vitally important that you have the correct training and instruction from the start. DSA holds an official voluntary register for LGV instructors. It has been set up in consultation with the training industry. To gain entry onto the register, instructors have to pass

- a test of their driving ability
- a test of their instructional ability

For further information about the voluntary register please telephone 0115 901 2500.

Books for study

You should already have a good, sound knowledge of driving skills It's strongly recommended that you study a copy of *The Highway Code*. You can buy one from any good bookshop or newsagent.

The DSA Driving Skills series of books will provide you with the information you need to further your skills and knowledge. *The Driving Manual* and *The Official Theory Test for Large Vehicle Drivers* are both particularly helpful.

There are other books and magazines produced by various publishers that can also give you information.

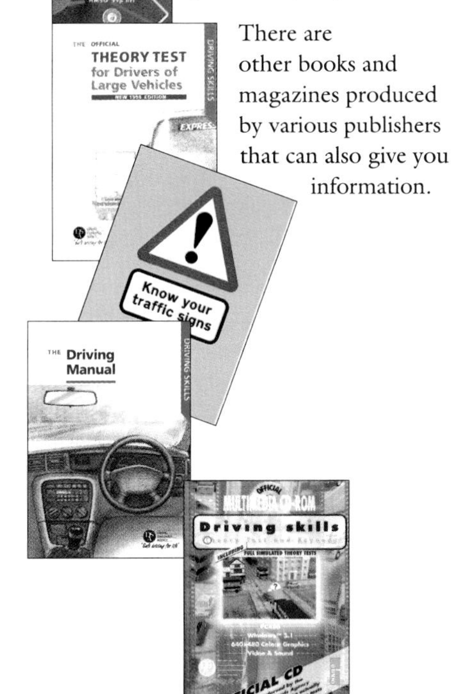

This part looks at the basic preparation needed to become a professional driver.

The topics covered

- Getting started
- The LGV driver
- Attitude
- Driving forces
- Vehicle characteristics

Applying for your licence

You should apply to the Driver and Vehicle and Licensing Agency (DVLA) in Swansea for the provisional entitlement to drive large goods vehicles (LGVs). An application form D1 is available from post offices.

In order to drive an LGV you must

- Have a full driving licence for a category B vehicle
- Hold a provisional LGV driving licence in the category that you wish to drive
- Meet the eyesight and medical requirements
- Normally be over 21 years old, unless you're a member of the armed forces authorised by the Ministry of Defence or a registered trainee within the 'Young Large Goods Vehicle (LGV) Driver Training Scheme'

Full details can be obtained from the DVLA enquiry line, 0179 277 2151, or from The Road Haulage and Distribution Training Council on 01908 313360.

If you passed your car driving test after 1 January 1997 and want to drive a vehicle between 3.5 and 7.5 tonnes you'll have to take a medium-sized lorry test (category C1) and meet the higher medical standards.

When you have your provisional licence you must

- Sign it

When driving as a learner you must

- Be accompanied by a qualified driver who has held a full licence for the category of vehicle being driven for at least 3 years*.
- Display L plates (or D plates, if you wish, when driving in Wales) to the front and rear of the vehicle

* Until April 2001, the accompanying driver need not have held the relevant licence for 3 years, as long as they held it since before 6 April 1998 and have held a full licence for another category within the same class (eg C1 for those wishing to supervise in category C) for the balance of the 3 year period.

You'll have to be fully qualified in a lower category of entitlement before seeking to gain entitlement in a higher category or sub-category. You'll have to

- Pass a category C test before taking a category C + E test
- Pass a category C or C1 test before taking a category C1 + E test

You won't have to gain a C1 before taking a test in category C.

Automatic transmission

If your vehicle doesn't have a clutch pedal it's classed as an 'automatic'. If you take the LGV driving test in an automatic vehicle your full LGV licence will restrict you to driving only LGVs fitted with automatic transmission.

Some modern vehicles have transmission systems where sensors select the next gear without the driver using the clutch pedal. These are also classed as 'automatic'.

Articulated vehicles

You must already hold a full licence to drive a rigid large goods vehicle (C1 or C) before you can apply for a provisional licence to drive an articulated vehicle (C1 + E or C + E). You don't have to pass a test in category C1 before taking a test in category C.

By then you'll already have experience of driving large vehicles. The information in this book about driving articulated vehicles will help you to prepare for your test and learn how to deal with the various characteristics of this type of vehicle.

If you're taking your test with a trailer you'll be expected to demonstrate uncoupling and recoupling during your test.

Medical requirements

Driving an LGV carries a heavy responsibility towards all other road users so it's vital that you meet exacting medical standards. For that reason your eyesight must meet much more stringent standards than that of ordinary drivers.

You may also be refused a licence to drive LGVs if you suffer from any serious medical condition that might cause road safety problems when driving such vehicles. An applicant or licence-holder failing to meet the epilepsy, diabetes or eyesight regulations must be refused in law.

It's your responsibility to notify immediately the Drivers Medical Unit at the DVLA, Swansea, if you have or develop any serious illness or disability that's likely to last more than three months and which could affect your driving.

The eyesight test

All drivers must be able to read, in good daylight, a number plate at 20.5 metres (about 67 feet). If you need to wear glasses or contact lenses to perform this test these must be worn while driving.

In addition, an applicant who hasn't previously held an LGV or PCV licence must by law have a visual acuity of at least

- 6/9 in the better eye
- 6/12 in the other eye

wearing glasses or contact lenses, if needed. She or he must also have an uncorrected visual acuity of 3/60 in both eyes.

Drivers who have held an LGV/PCV licence before 1 January 1997 but who don't meet the new standard may still qualify for a licence.

If you require any further general information you should contact the DVLA on 0179 277 2151. For enquiries about medical standards you should contact the

Drivers Medical Unit
DVLA
Swansea
SA99 1TU

If you normally wear glasses or contact lenses, always wear them whenever you drive.

Medical examination and form D4

Consult your doctor first if you have any doubts about your fitness. In any case, if this is your first application for LGV entitlement a medical report must be completed by a doctor. You'll also need to send in a medical report with your application if you're renewing your LGV licence and you're aged 45 or over, unless you've already sent one during the last 12 months.

You'll need to have a medical examination in order to complete form D4.

Only complete the applicant details and declaration (Section 8 on the form) when you're with your doctor at the time of the examination. Your doctor will complete the other sections. The medical report will cover

- Vision
- Nervous system
- Diabetes melitus
- Psychiatric illness
- General health
- Cardiac health
- Medical practitioner details

Study the notes on pages 1 and 2 of form D4 then remove these two pages before sending in your application. Keep those pages for any future reference.

This medical report isn't available free under National Health rules. Your doctor is entitled to charge the current fee for this report. You're responsible for paying this fee: it can't be recovered from the DVLA. In addition, the fee isn't refundable if your application is refused.

The completed form must be received by the DVLA within four months of the date of your doctor's signature.

Medical standards

You may be refused an LGV driving licence if you suffer from any of the following

- Liability to epilepsy/seizure
- Diabetes requiring insulin (unless you held a licence on 1 April 1991 and the Traffic Commissioner who issued that licence had knowledge of your condition)
- Visual defects (see the eyesight requirements on p. 9)
- Heart disorders
- Persistent high blood pressure (see the notes on form D4 for details)
- Strokes/unconscious lapses within the last five years
- Any disorder causing vertigo within the last two years
- Severe head injury, with serious continuing after-effects, or major brain surgery
- Parkinson's disease, multiple sclerosis or other chronic nervous disorders likely to affect the use of the limbs
- Mental disorders
- Alcohol/drug problems
- Serious difficulty in communicating by telephone in an emergency
- Visual field defects

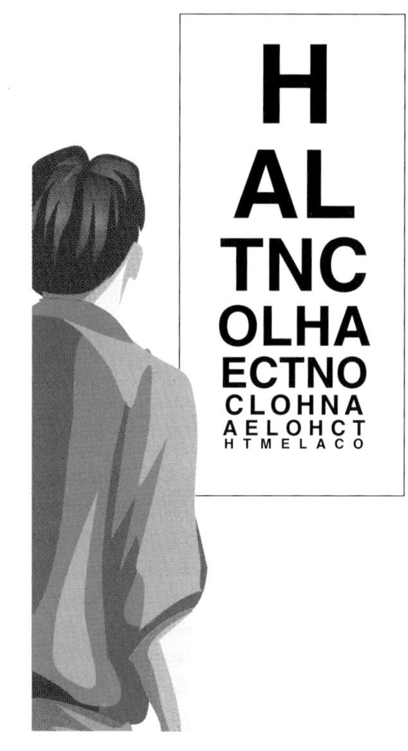

Professional standards

Driving an LGV requires skill combined with knowledge and the right attitude and driving techniques.

To become a professional driver you'll need

- The demanding driving skills required
- The knowledge to deal with all the regulations that apply to your work
- A comprehensive knowledge of *The Highway Code* including the meaning of traffic signs and road markings, especially those which indicate a restriction for LGVs

From the start, you'll need to appreciate the differences between driving small and large goods vehicles. It's also essential to understand the forces at work on your vehicle and its load.

Initially, the most important thing to learn is that the *way* you drive is vital.

- Drive properly and safely and the goods entrusted into your care will arrive safely at their destination
- Drive dangerously or even carelessly and the potential for disaster is enormous

Whether you're driving an unladen lorry at the 7.5 tonne end of the scale or a fully laden articulated vehicle of 38 tonnes or more, you can never act hastily without serious consequences resulting.

A loaded LGV travelling at speed and colliding with another vehicle will cause serious damage. You're the one with the responsibility of driving your vehicle safely at all times.

Your vehicle will probably have the owners' name on display, therefore your driving will be similarly on display. Make sure that your vehicle is clean and well maintained. Your driving should match the same high standards. Show a good example of skill, courtesy and tolerance to other road users. Be a credit to yourself, your company and your profession.

Your LGV licence is a privilege which requires effort to gain and even more effort to keep.

Appropriate behaviour

As a professional driver you should set a good example of driving to others. You should always have an idea of how other road users see you. Be aware that they might not understand why you take up certain positions to make turns or take longer to manoeuvre.

You'll spend a great deal of time at the wheel of your vehicle. Losing your temper or having a bad attitude towards other road users won't make your working life pleasant. A good attitude will help you to enjoy your work and is safer for others around you.

Tailgating

The sheer size, noise and appearance of a typical LGV often appears somewhat intimidating to a cyclist, motorcyclist or even the average car driver. Travelling dangerously close behind a smaller vehicle at speed can be very intimidating for the vehicle in front. When an LGV appears to be being driven in an aggressive way other road users often feel really threatened.

If you drive too close to the vehicle in front your view of the road ahead will be severely restricted. You won't be able to see or plan for any hazards that might occur. The room in which you have to stop is also reduced, probably within the stopping distance for the speed at which you're travelling. This is a dangerous practice.

Police forces are concerned at the number of accidents that have been caused as a direct result of vehicles driving much too close to each other. A number have mounted campaigns to video and prosecute offenders.

In an effort to improve the image of the transport industry some large retail organisations are reviewing the placing of contracts with any distributor whose vehicles have been seen repeatedly tailgating on motorways, etc.

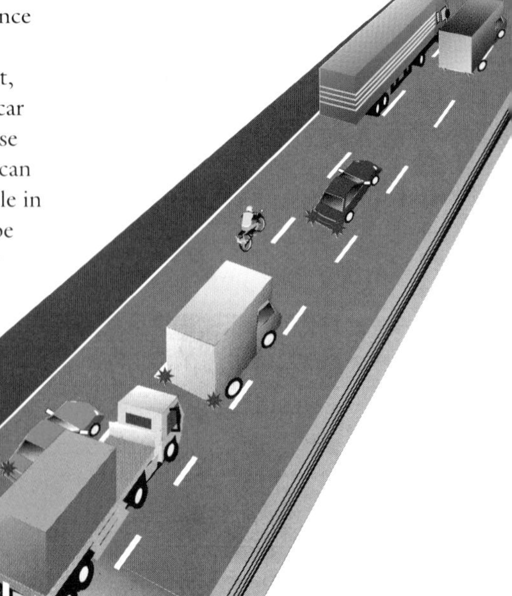

Intimidation

Don't use the size, weight and power of your vehicle to intimidate other road users. Even the repeated 'hiss' from air brakes being applied while stationary gives the impression of 'breathing down the neck' of the driver in front.

Speed

The introduction of 'just in time' flow-line policies reduces the need for manufacturers to hold large stocks of materials. These policies are also intended to ensure the delivery of fresh foods at the supermarket, for example.

Don't allow your employer to set unachievable delivery targets. You should never be under extreme pressure to meet deadlines.

You can never justify driving too fast simply because you need to reach a given location by a specific time, whether it be a ferry, loading bay or depot.

If an accident results and you injure someone there's no possible defence for your actions.

Retaliation

You should resist at all times impatience or the temptation to retaliate. Always drive

- Courteously
- With anticipation
- Calmly, allowing for other road users' mistakes
- With full control of your vehicle

You can't act hastily without the possibility of serious loss of control when driving an LGV.

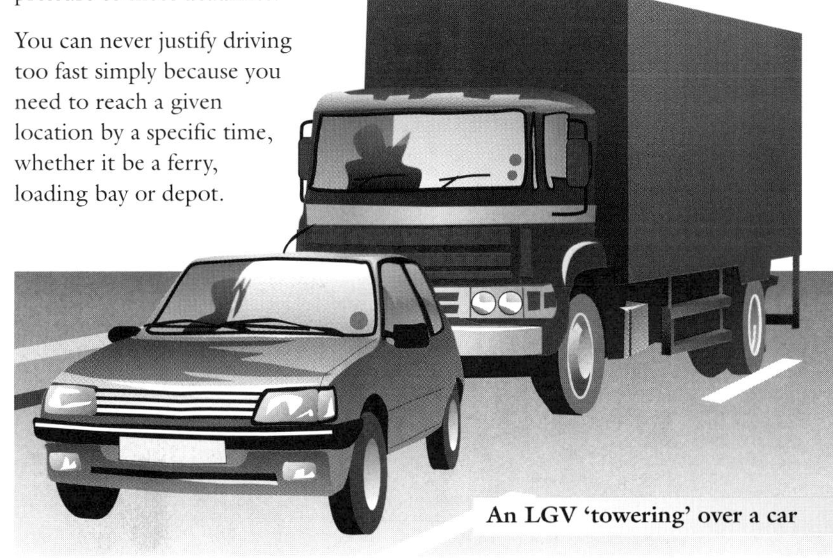

An LGV 'towering' over a car

The horn

Because LGVs are often equipped with powerful multi-tone air horns their use should be strictly confined to the guidance set out in *The Highway Code* – to warn other road users of your presence.

Don't use the horn

- Aggressively
- Between 11.30 pm and 7 am unless a moving vehicle poses a danger

The headlights

To avoid dazzle don't switch the headlights on to full beam when following another vehicle. Don't

- Switch on additional auxiliary lights that may be fitted to your vehicle unless the weather conditions require their use
- Repeatedly flash the headlights while driving directly behind another vehicle

Flashing the headlights lets other road users know that you're there. It doesn't mean that you wish to give up or show priority. You may be misunderstood by others when using an unauthorised code of headlight flashing, which could lead to accidents.

Neither the headlights nor the horn(s) must be used to rebuke or to intimidate another road user. Courtesy and consideration are the hallmarks of a professional driver.

Effects of your vehicle

As a competent LGV driver you must always be aware of the effect your vehicle and your driving has on other road users.

You need to recognise the effects of turbulence or buffeting caused by your vehicle, especially when passing

- Cyclists
- Motorcyclists
- Cars
- Cars towing caravans
- Other lorries and buses
- Horse riders
- Pedestrians

On congested built-up roads, particularly in shopping areas, take extra care when you need to drive closer to the kerb. Be aware of

- The possibility of a pedestrian stepping off the kerb (and under the wheels)
- The nearside mirror striking the head of a pedestrian standing at the edge of the kerb
- Cyclists moving up the nearside of your vehicle in slow-moving traffic

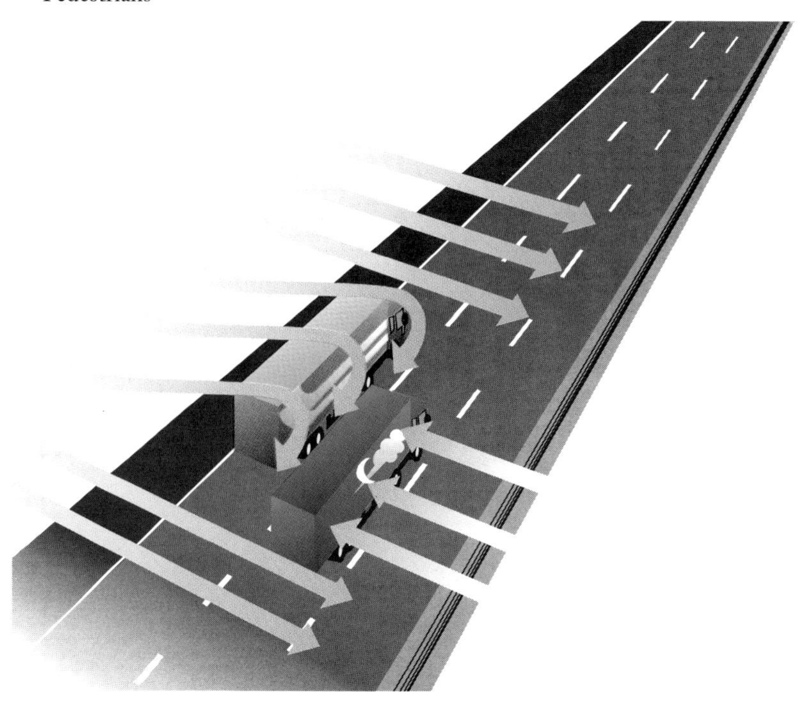

Cyclists

Over a quarter of cyclist deaths are as a result of collisions with LGVs. You need to be aware of the limited vision you have around your vehicle due to its size and shape. Use your mirrors so that you have a constant picture of what's happening all around. Always check any blind spots before you move away.

Also remember that cyclists could become unbalanced by the buffeting effect of a large vehicle passing closely.

Understanding LGVs

You'll need to study and understand the information given in this book before you can consider taking an LGV driving test. You also need to understand how various kinds of LGVs handle in order to drive them safely.

To drive a lorry safely you must first appreciate the main differences between driving small and large goods vehicles. These are

- Weight
- Width
- Length
- Height
- Distance needed to pull up
- Distance needed to overtake
- Control needed when going downhill
- Power needed to climb uphill
- The need to avoid any sudden changes of speed or direction

Some of these aspects will be obvious from the moment you first start to drive an LGV. Other features will only become apparent after you've gained experience. The essential factor is to recognise that much more forward planning and anticipation is needed to drive an LGV safely.

Whether an LGV is

- Laden or unladen
- Rigid
- Towing a drawbar trailer
- Articulated

it's most stable when travelling in a straight line under gentle acceleration. Sudden or violent

- Steering
- Acceleration
- Braking

can cause severe loss of control. All braking should be carried out smoothly and in good time.

Most modern large vehicles are fitted with an air braking system. Older vehicles might have a hydraulic braking system. In *this* case, if you find you need to 'pump' the brakes STOP as safely as you can in a convenient place and check the hydraulic system. Don't drive on unless you're sure that you can stop safely.

Erratic handling

Sudden acceleration forward might cause an insecure load to fall off the back of a vehicle. Similarly, if harsh braking is applied

- The load may attempt to continue moving forward
- The tyres may lose their grip on the road surface causing the vehicle to skid
- The weight of the vehicle is transferred forward causing the front of the vehicle to 'dip' downward

Any sudden steering movement may also unsettle the load and cause it to move. Any movement of the load is likely to make the vehicle unstable.

All acceleration and braking should be controlled and as smooth and progressive as possible.

Friction

The grip between two surfaces is called friction. The grip that rubber tyres have on a road surface produces traction (force), which is essential when

- Moving away or accelerating
- Turning/changing direction
- Braking/slowing down

The amount of traction will depend on

- The weight of the vehicle
- The vehicle's speed
- The condition of the tyre tread
- The tyre pressure
- The type and condition of the road surface
  – anti-skid
  – loose
  – smooth
- Weather conditions
- Any other material present on the road
  – mud
  – wet leaves
  – diesel spillage
  – other slippery spillages
  – inset metal rail lines
- The rate of change of speed or direction (sudden steering/braking)
- The condition of mechanical components
  – steering alignment
  – suspension

Sudden acceleration or deceleration can cut down the grip your vehicle has on the road. Under these conditions the vehicle will

- Lose traction (wheelspin)
- Break away on a turn (skid)
- Not stop safely (skid)
- Overturn

Jack-knifing

This is usually more likely to occur with an unladen vehicle.

In the case of an articulated vehicle, severe braking can result in jack-knifing as the tractive unit is pushed by the semi-trailer pivoting around the coupling (fifth wheel). This is even more likely if the vehicle isn't travelling in a straight line when the brakes are applied.

Similar results will occur with a drawbar trailer, where there may be **two** pivoting points: at the coupling pin and at the turntable of the front wheels of a two- or more axle trailer.

Changing into a lower gear when travelling at too high a speed or releasing the clutch suddenly will produce much the same effect. That is, the braking effect will only be applied to the driven wheels.

For some years now, well-proven systems have been available that reduce the risk of jack-knifing on articulated vehicles.

Some vehicles are equipped with retarders, a sophisticated system of electronic signals that ensures appropriate braking effort is applied to the trailer wheels.

Trailer swing

This can occur on a drawbar combination (or occasionally on an articulated unit) when

- Sharp braking is applied on a bend
- Excessive steering takes place at speed
- The brakes on either the tractive unit or trailer aren't properly adjusted

It follows that all braking, gear changes, steering and acceleration should be smooth and under full control.

Forces affecting your vehicle

Gravity

When a vehicle is stationary on level ground the only force generally acting upon it (ignoring wind forces, etc.) is the downward pull of gravity.

On an uphill gradient gravity will have a greater effect on a moving vehicle and its load so that

- More power is needed from the engine to move the vehicle and its load forward and upward
- Less braking effort is needed and the vehicle will pull up in a shorter distance

On a downhill gradient the effect of gravity will tend to

- Make the vehicle's speed increase
- Require more braking effort
- Increase stopping distances

A vehicle's centre of gravity is the point around which all of its weight is balanced. To keep the vehicle and its load stable this should be arranged to be

- As low as possible
- Along a line running centrally down the length of the LGV

The higher this centre of gravity occurs, the less stable a vehicle and/or its load will be. As a result the vehicle will become more easily affected by

- Braking
- Steering
- The slope (camber) of a road
- The wheels running over a kerb and resulting in either the load tilting or falling, or the vehicle overturning

End-tipper vehicles When a loaded tipper vehicle body (whether tanker, bulk carrier or high-sided open body) is raised to discharge a load the centre of gravity is raised to a critical position. It's vitally important to ensure that the vehicle is on a level, solid surface before engaging the hoist mechanism.

Ensure that there are no overhead power lines in the vicinity. Keep clear of scaffolding or any other obstructions.

Side-tipper vehicles Always select the firmest level site available before tilting the vehicle body. Until the load is discharged all the weight will be transferred to one side. Unless the vehicle is on firm level ground there is a risk of it overturning.

Always take the time to check the ground before tipping. Get out and check all around the vehicle. Make sure that it's safe before you tip.

Kinetic energy

This is the energy contained in a moving vehicle. The amount depends on the

- Mass (weight) of the vehicle + load
- Speed

Kinetic energy must be reduced by the brakes in order to stop a vehicle. The kinetic energy of a stationary vehicle is zero.

An increase in speed from 15 mph to 45 mph (x 3) increases the kinetic energy **ninefold** (= 3 x 3). If you reduce the speed by half, say from 50 mph to 25 mph, the kinetic energy acting on your vehicle is **one-quarter** of what it was before braking.

As the brakes reduce the speed of a vehicle, kinetic energy is converted into heat. Continuous use of the brakes can result in them becoming overheated and losing their effectiveness, especially on long downhill gradients. This is known as 'brake fade'.

The effort required to stop a fully laden LGV travelling at 56 mph (85 kph) is so much greater than that needed to stop an ordinary motor car travelling at similar speed. You'll need to allow extra time and space to stop an LGV safely. Don't follow other vehicles too closely – always leave a safe separation distance. In addition, harsh braking should be avoided at all times when driving an LGV.

Momentum

This is the tendency for a vehicle and/or its load to continue in a straight line. It depends on

- Mass (weight)
- Speed

The higher the speed, the greater the momentum and the greater the effort required to

- Stop
- Change direction

Forces on the load If the forces acting on a load cause it to become detached from the vehicle the load will move in the direction of the force. While

- Accelerating it will fall off the back
- Braking it will continue moving forward
- Tilting it will topple over
- Turning it will continue on the original path and fall off the side

Centrifugal force

When a vehicle takes a curved path at a bend the force acting upon it will attempt to cause the vehicle to continue on the original straight course. At low speeds this force will be overcome by the traction of the tyres on the road surface.

If a loaded vehicle takes a bend at too high a speed the centrifugal force will act on the vehicle and may also cause the load to become detached and fall off the side.

Loss of control

If you ask too much of your tyres by turning and braking at the same time you're dividing the available traction. Once any or all the tyres lift or slide you're no longer in control of the vehicle. What happens next will depend on the particular forces that are acting on the vehicle.

When any change is made to a vehicle's motion or direction the same forces will act on any load being carried. Unless a load is secured to the vehicle so that it can't move and the traction between the tyres and the road surface is maintained you're going to lose control of your vehicle.

Shedding loads

There are several differences between driving a large vehicle and driving a smaller vehicle or a car. An unladen vehicle will handle differently from a laden one. Handling will also change depending on the size or type of load carried.

The causes of shed loads are

- Driving error
  - sudden change of speed or direction
  - too high a speed
  - skidding
- Instability of load
  - unsuitable vehicle
  - badly stowed
  - movement in load
  - restraints failed
  - unsuitable type of restraint
- Mechanical failure
  - suspension failure
  - tyre failure
  - trailer disengaged
  - wheel loss
- Collision
  - another vehicle
  - a bridge, etc.
  - lamp-posts
  - signs
  - signals
  - bollards

Most of these situations are preventable.

You should know the handling characteristics of the vehicle that you're driving and drive in a safe and sensible manner.

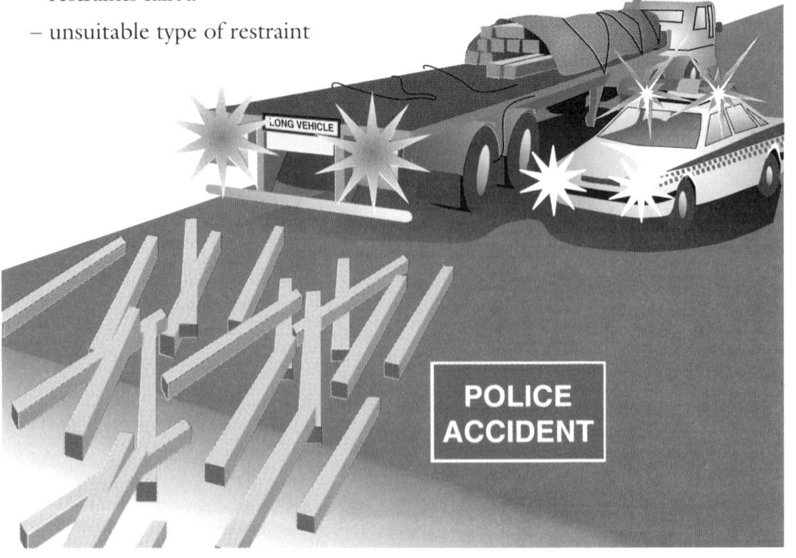

Maintaining control

You can't alter the severity of a bend or change the weight of a load. However, you have control over the **speed** and **braking** of your vehicle.

Reduce speed in good time before negotiating

- Bends
- Roundabouts
- Corners

To keep control you should ensure that all braking is

- Controlled
- In good time
- Made when travelling in a straight line, wherever possible

Avoid braking and turning at the same time (unless manoeuvring at low speed). Look **well** ahead to assess and plan.

The different types of large goods vehicles will each require specific handling. You'll need to bear this in mind if you want to become a professional LGV driver.

Short wheel-base vehicles

- Will bounce more noticeably than some long wheel-base vehicles when empty. This can affect braking efficiency and all-round control
- Shouldn't be pushed into bends or corners at higher speeds simply because the vehicle appears to be easier to drive

Long wheel-base rigid vehicles

These require additional room to manoeuvre, especially

- When turning left or right
- Negotiating roundabouts
- Entering or leaving premises

Typical examples of this type of vehicle are

- Removal vans
- Box vans
- Brick carriers
- Bulk carriers (for aggregates, etc.)
- 'Eight-wheeler' tankers

Box vans

In addition to the extra space needed when turning, the box type of body, when lightly loaded or empty, is very susceptible to crosswinds on exposed stretches of road. Failure to heed adverse weather warnings may well result in the vehicle being blown over. For this reason, high-sided vehicles are often banned from using certain roads and bridges where these problems are known to occur.

It's essential to observe any temporary speed limits imposed during these conditions.

Articulated vehicles

Drivers of these vehicles should be especially vigilant before turning corners or negotiating roundabouts.

Failure to plan ahead and select the correct course might result in the rear wheels 'cutting in'. This could cause your vehicle to

- Clip the kerb
- Collide with street furniture

It could also endanger

- Pedestrians
- Cyclists
- Other vehicles

You should avoid overshooting left or right turns.

The stability of the vehicle is at risk if an excessive steering lock is applied in order to make a 'swan-neck' turn.

Articulated tankers

The vehicle wheels on the inside of a curve may start to lift if the

- Centre of gravity is high enough
- Speed is high enough
- Vehicle is travelling sufficiently fast
- Vehicle is driven on a curved path

In many cases this is followed seconds later by 'roll-over'.

The problem frequently involves tanker vehicles carrying fluids in bulk. The reasons for this problem have been attributed to a number of possible causes.

It appears most likely to occur when modern, heavier and more powerful vehicles (and, in the majority of cases, those equipped with power steering) reach a critical situation. As an example, the forces acting on a typical loaded articulated tanker vehicle negotiating a roundabout at a speed of about 25 mph will cause it to overturn if only a further quarter turn is applied to the steering wheel.

You must be fully alert to the possibility of this happening. Adjust the speed of your vehicle to avoid wheel-lift and roll-over.

'Roll-over' The type of suspension fitted to a vehicle will influence its resistance to roll-over.

Modern tri-axle semi-trailers fitted with single wheels on each side have extended the tracking width available, compared to twin-wheeled units, and improved resistance to roll-over.

Although vehicles equipped with air suspension systems are often considered to possess improved anti-roll stability compared to traditional steel-leaf spring suspension, test results have shown that there are similarities in their level of stability.

The transition from wheel-lift to roll-over is much more rapid on vehicles equipped with air suspension systems, giving very little warning that this is about to take place.

Large petrochemical companies include specialised driver training to avoid such incidents happening.

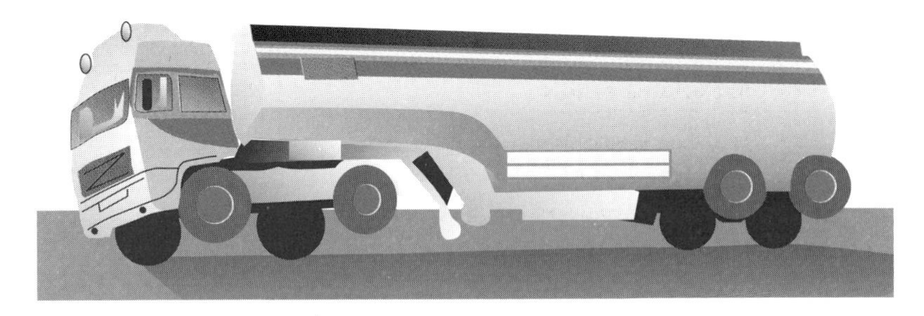

The wave effect If the drivers of certain tanker vehicles relax the footbrake when braking to a stop there's a danger that the motion in the fluid load could force their vehicles forward. This is due to the wave effect created in the tank contents, especially where baffle plates are omitted from the tank design. (Such tanks are sometimes found on vehicles carrying foodstuffs or where cleaning out the tanks presents any difficulties.)

Using walkways Drivers of tanker vehicles must exercise special care when climbing onto walkways to gain access to tank hatches. Not only is this to avoid injury as a result of slipping off, but also to avoid the danger of overhead cables, pipeways, etc.

Venting All tanks must be vented according to instructions. This will avoid serious damage to the tanker body as the external air pressure becomes greater than the pressure within the tank.

Compressed gases Drivers of vehicles carrying compressed gases, especially at low temperatures (e.g., liquid nitrogen, oxygen, etc.), must comply with regulations relating to the transport of such materials.

Fire or explosion In the case of hazardous materials, all safety precautions must be strictly followed, especially where there's a risk of fire or explosion.

The electrical systems of vehicles carrying petrochemicals and other highly inflammable materials are amended to meet stringent safety requirements. No unauthorised additions or alterations must be made to such vehicles. Any defects must be reported immediately.

The appropriate fire-fighting equipment must be available and drivers trained in its use.

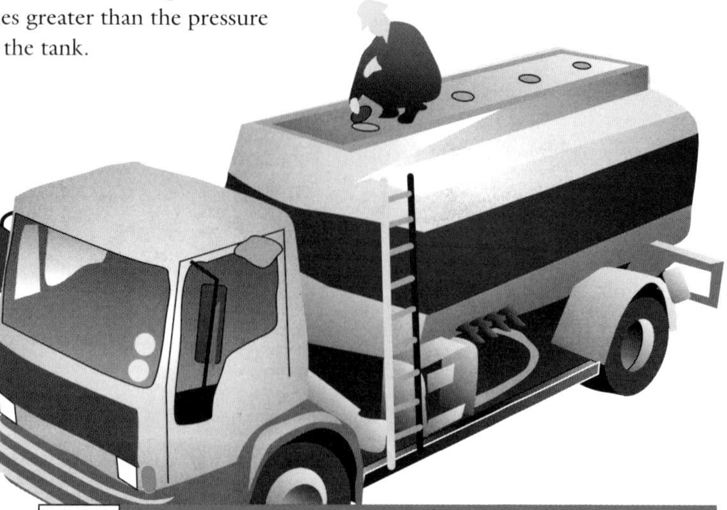

Articulated car transporters

These vehicles require even higher standards of driving and anticipation by the driver.

The overhang created by the top deck swings through a greater arc than the cab of the tractor unit, particularly when negotiating turns. This means that there's a risk of collision with

- Traffic signals
- Lamp-posts on central refuges
- Traffic signs
- Walls and buildings

You should plan ahead and take an appropriate avoiding line on approach to turns when driving these vehicles.

The stability of these vehicles also needs to be considered. It's a case of last on, first off, so the lower deck may be clear of vehicles while there may be several still on top. The centre of gravity is substantially shifted in such cases.

You should be aware of the height of your vehicle at all times, especially if you're carrying vans.

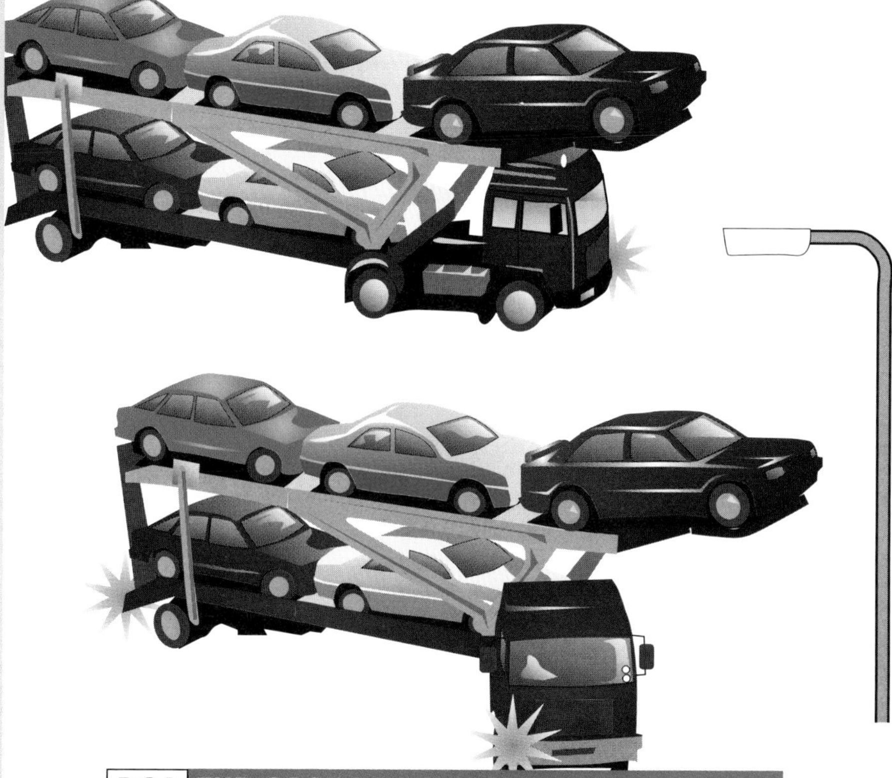

Demountable bodies

This type of vehicle is similar to containers except that the body is fitted with 'legs', which can be lowered to enable the carrier vehicle to drive under or out.

Care needs to be taken to ensure that the

- Legs are secured in the down position before the vehicle is driven out
- Legs are secured in the up position before the vehicle moves off
- Height is correct and the body is stable before driving the carrier vehicle underneath
- Surface is firm and level before 'demounting' the body

Another type (often an open, bulk high-sided body) is fitted with skids and is winched up onto the carrying vehicle. Apart from the dangers of overloading, care must be taken when operating such winches.

The construction of these vehicles often means that the centre of gravity can be higher than normal when conveying a loaded skip. Take this into account, especially when cornering.

Double-decked bodies

These vehicles are constructed to give increased carrying capacity to box van or curtain-side bodies. They're frequently used in the garment distribution sector.

Care must be taken to ensure that when the vehicle is in transit the lower deck isn't left empty while a load remains on the top tier. Such a shift in the centre of gravity will result in the high-sided vehicle being even more vulnerable in high winds and therefore more likely to overturn.

Refrigerated vehicles

When driving refrigerated or 'reefer' vehicles carrying suspended meat carcasses, care must be taken to avoid the 'pendulum' effect when cornering. Always reduce speed in good time.

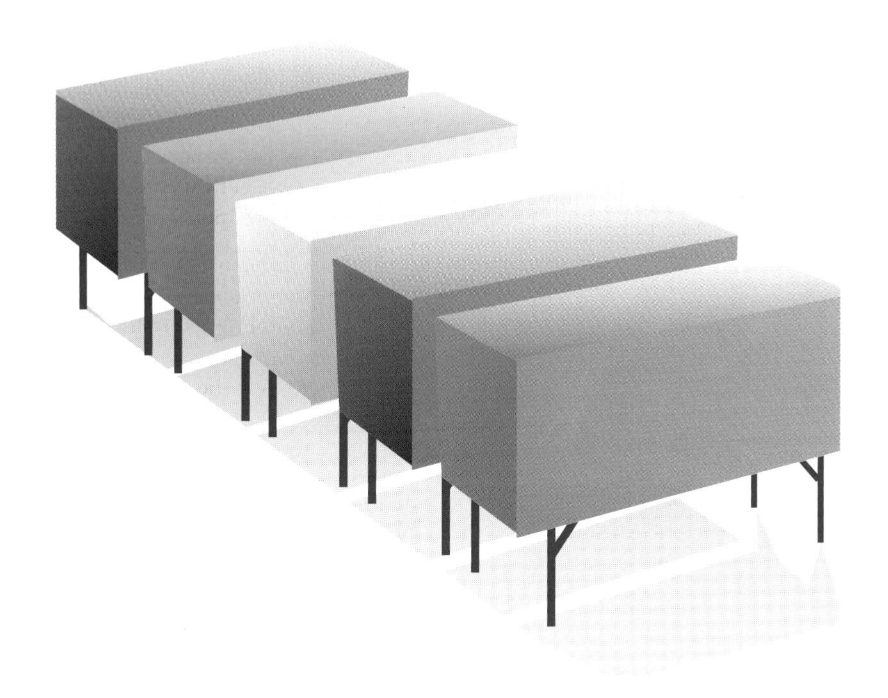

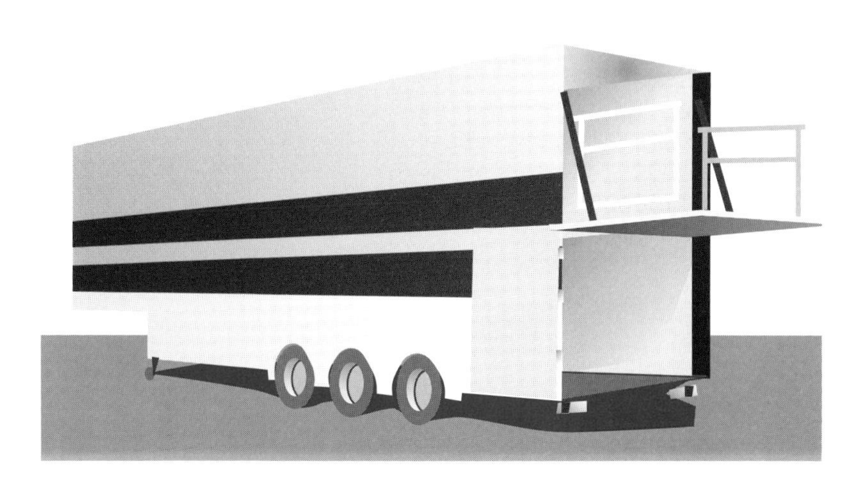

PART TWO LIMITS AND REGULATIONS

This part looks in detail at the limits, rules and regulations that you'll need to know and follow when you become a professional LGV driver.

The topics covered

- Vehicle limits
- Braking systems
- Load restraint
- Environmental impact
- Legal requirements
- Other regulations

Basic knowledge

The transport industry is subject to an extremely large number of regulations and requirements relating to

- Drivers
- Operators
- Companies
- Vehicles
- Goods

It's essential that you keep up to date with all the changes in road transport legislation that affect you.

The first thing you'll need to know about is your vehicle. The various aspects to consider are its

- Weight (restrictions)
- Height (clearances, etc.)
- Width (restrictions)
- Length (clearances)
- Ground clearance (low-loader or dual-purpose trailers only)

You'll also need to know the speed limits that apply to your vehicle and the speed at which it will normally travel.

Weight

It's essential that you're aware of, and understand the limits relating to, any vehicle that you drive. (Definitions of terms relating to weight limits can be found in the glossary of terms in Part Six.)

Weight limits in many cases refer to the maximum gross weight (MGW, or maximum authorised mass, MAM). You can drive heavier vehicles if engaged in inter-modal operations between road and rail terminals. These are subject to a number of conditions

- The number of axles
- Tyre specifications
- Axle spacings
- Road-friendly suspension

Also, special documentation is required to cover such movements between rail-head and road depot.

Road-friendly suspension generally means that an axle is fitted with air suspension, or a suspension regarded as being equivalent to air suspension (EC directive 92/7/EEC).

At least 75% of the spring effect is caused by an air spring (that is, operated by air or some other compressible fluid). If this sounds complicated it's intended to illustrate that the law relating to weight limits *is* complex.

Accurate information is required to ensure that the vehicle complies with the law and also that the driver isn't committing any offence.

New technology allows local authorities to make spot weight checks on vehicles. Roadside checks frequently reveal contraventions of many weight limit regulations. Once identified, the police may escort a vehicle and its driver to the nearest official weighbridge.

It's essential that all limits are complied with in order to avoid overloading, and possible prosecution.

Weight restrictions normally apply to the plated weight of a vehicle. When tractor units of articulated LGVs are being driven without a trailer they're still subject to weight limits. These relate to

- Lighting regulations
- Loads subject to weight limits
- Lanes from which LGVs are banned on multi-lane motorways

Regulations also include axle weight limits applying to the tractor unit, which must not be exceeded when the vehicle is loaded.

It's essential that you ensure that the load is distributed correctly and safely, either on or in your vehicle. This means that loads are built up against the headboard or front wall of the body in order to reduce movement in transit. When loading your vehicle ensure that the front axle(s) isn't/aren't overloaded.

Articulated units and part-loads

To increase stability and reduce the risk of the trailer wheels lifting when turning it's preferable to have part-loads (such as an empty single ISO* container) located over the rear axle(s).

Whatever goods vehicle you drive you should study the Department of Transport publication *Code of Practice: Safety of Loads on Vehicles.*

*ISO = International Standards Organisation

Recovery vehicles

Always make sure that axle loading limits aren't exceeded when a recovery vehicle removes another LGV by means of a suspended tow. This is sometimes overlooked.

Height

Legislation on travelling height is expected to be introduced in 1997, which will require the following.

If you're the driver of any vehicle where the overall travelling height of the vehicle, its equipment and load (including any trailer) is more than 3 metres (10 feet) you should ensure that

- The overall travelling height is conspicuously marked in figures not less than 40 mm high in such a manner that it can be read from the driving position. It should be marked in feet + inches, or in feet + inches and in metres
- Any height indicated isn't less than the overall travelling height of the vehicle

A height notice isn't required on a particular journey if you're carrying sufficient information in documents about the route or choice of routes. This should include the height of bridges and other overhead structures to allow you to complete your journey without any risk of a collision with an overhead structure. You must then travel on a route described in these documents.

Warning devices

If a vehicle, and any trailer drawn by it, is a maximum height of more than 3 metres (10 feet) a visual warning device should warn the driver if the highest point of the equipment exceeds a predetermined height when the vehicle is being driven.

The predetermined height must not exceed the overall travelling height by more than 1 metre (3 feet) unless the equipment has a mechanical locking device that can lock it in a stowed position. In addition, the equipment needs to be fixed in that position by the locking device when the vehicle is being driven.

The requirement for height notices, route information or warning devices won't apply if on a particular journey you're highly unlikely to encounter any bridge or overhead structure which isn't at least 1 metre (3 feet) higher than the maximum or overall travelling height of the vehicle.

Overhead clearances

Drivers of any vehicle exceeding 3 metres (10 feet) in height should exercise care when entering roofed premises such as

- Loading bays
- Depots
- Dock areas
- Freight terminals
- Service station forecourts
- Any premises that have overhanging canopies

or when negotiating

- Bridges
- Overhead cables
- Overhead pipelines
- Overhead walkways
- Road tunnels

Over-height loads

If your load exceeds 5.25 metres (17 feet 6 in.) notification is required to be sent to telephone companies.

Electricity Supply Regulations 1988 state that the minimum height of electricity supply cables over roads must be at least 5.8 metres (19 feet), but any vehicle or load that exceeds 5 metres (16 feet 6 in.) is at risk. Contact the local electricity board for advice regarding the advance notice to be given to the electricity authorities concerned (at least 19 days). You should inform them of the load, routes, etc. See your local telephone directory for details.

When planning the movement of such vehicles and loads the local Highways Authority will need to be contacted where overhead gantry traffic signs or suspended traffic lights are likely to be affected by loads over 5 metres (16 feet 6 in.).

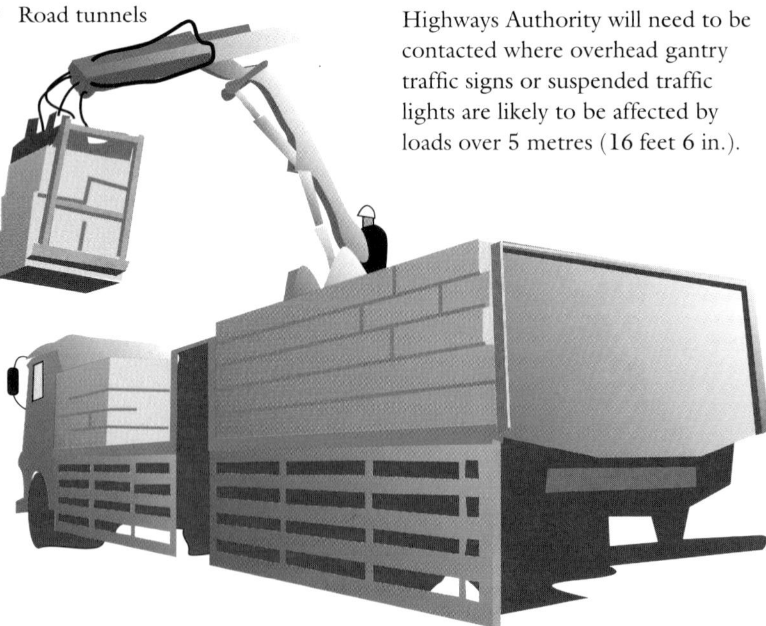

Bridges

Every year around 800 incidents occur involving vehicles or their loads hitting railway or motorway bridges. An impact with any bridge can have serious consequences. An LGV, or any part of its load, colliding with a railway bridge could result in the bridge being weakened. This could also inconvenience rail traffic or create the potential for a major disaster. These factors are in addition to the costs involved in making the bridge safe, re-aligning railway tracks, and the general disruption to road traffic.

The headroom under bridges in the UK is at least 5 metres (16 feet 6 in.) unless marked otherwise. But remember, this might mean the maximum height at only the **highest** point of an arch.

If your vehicle collides with a bridge you must report the incident to the police. If a railway bridge is involved you should also report it to Railtrack on 0345 003 355 immediately.

Do this immediately.

Give information about

- The location
- The damage
- Any bridge reference number (often found on a plate bolted to the bridge or wall)

You must also inform the police within 24 hours, if you don't do so at the time of the incident. Failure to notify the police is an offence.

Know your height

You **must** know the height of your vehicle and its load. Don't guess. If in doubt, measure it. Also, don't ignore

- Traffic signs
- Road markings
- Warning lights

Don't take chances.

If you aren't sure of the safe height STOP and call the authorities. Always

- Plan your route
- Slow down when approaching bridges
- Know the overall vehicle/load height
- Keep to the centre of arched bridges
- Wait for a safe gap to proceed if there's oncoming traffic

Height Guide

Metres	Feet/Inches	
5.0	16	6
4.8	16	
4.5	15	
4.2	14	
3.9	13	
3.6	12	
3.3	11	
3.0	10	
2.7	9	

15'3"
4.6m

WIDE LOAD

Width

As the driver of an LGV you must be aware of the road space that the vehicle occupies. This is particularly important where width is restricted because of parked or oncoming vehicles. Look out for signs showing restrictions.

Traffic calming measures are becoming much more common. Don't get into a situation where you're forced to reverse or turn due to width restrictions.

Overall Width Limits

Metres	Ft	In	
2.75	9	4	Locomotives
2.6	8	5	Refrigerated vehicles and trailers*
2.5	8	2	Motor tractors
2.5	8	2	Motor cars
2.5	8	2	Heavy motor cars
2.5	8	2	Trailers*
2.3	7	6	Other trailers

* Subject to certain conditions

Wide loads

Loads projecting over 305 mm beyond the width of the vehicle or those over 2.9 metres (9 feet 5 in.) but less than 3.5 metres (11 feet 5 in.) require side markers and notification to the police.

Wide loads that are over 3.5 metres (11 feet 5 in.) but less than 4.3 metres (14 feet 1 in.) require side markers, police notification and an attendant.

Wide loads that are 4.3 metres (14 feet 1 in.) to 5 metres (16 feet 6 in.) are also subject to the following speed limits

- 30 mph on motorways
- 25 mph on dual carriageways
- 20 mph on all other roads

Wide loads between 5 metres (16 feet 4 in.) and 6.1 metres (20 feet) require

- Side markers
- Police notification
- An attendant
- Department of Transport approval

and are subject to the above speed limits.

Side marker boards must comply with regulations so that they show clearly on either side of the projection to the front and to the rear. All marker boards must be independently lit at night.

Length

Locations where length restrictions apply are comparatively few but they include

- Road tunnels
- Level crossings
- Ferries
- Certain areas in cities

Drivers of long rigid vehicles, either with or without drawbar trailers, or articulated LGVs must be aware of the length of their vehicle especially when

- Turning left or right
- Negotiating roundabouts or mini-roundabouts
- Emerging from premises or exits
- Overtaking
- Parking, especially in lay-bys
- Driving on narrow roads where there are passing places
- Negotiating level crossings

Maximum Length Limits

Metres	Ft/In	
12	39 4	Rigid vehicles
16.5	54	Articulated vehicles*
18 or 18.35	59 60	Vehicle and trailer combinations**
18	59	Articulated vehicles with low-loader semi-trailer manufactured after 1 April 1991 (not including step-frame low-loaders)

Car transporter semi-trailers

Metres	Ft/In	
12.5	41	Kingpin to rear
4.19	13 8	Kingpin to any point at the front

Other semi-trailers

Metres	Ft/In	
12	39 4	Kingpin to the rear
2.04	6 7	Kingpin to any point at the front
14.04	46	Composite trailer
12	39 4	Drawbar trailer with four or more wheels, and drawing vehicle is more than 3,500kg MGW
7	23	Other drawbar trailers

* Maximum length limit for vehicles

Applying the brakes

Planning and anticipation of road hazards should remove the need for harsh braking. Harsh or heavy braking can result in a vehicle's wheels locking, leading to a loss of vehicle control particularly on slippery road surfaces (for example, wet, icy or snow-covered).

In an emergency braking situation you may need to brake heavily. If your vehicle is equipped with an anti-lock braking system (ABS)* and you're unlikely to be able to stop the vehicle before reaching an obstruction, apply maximum force to the brake pedal, maintaining this force. You shouldn't 'pump' the brake pedal as this will reduce the effectiveness of the ABS system. If your vehicle doesn't have an ABS, 'wheel lock' can be controlled during heavy decelerations by 'cadence' braking, that is, rapid pumping of the brake pedal.

*ABS is the registered trade mark of Bosch (Germany) for Anti Blockiersystem.

Types of brakes

There are three types of braking systems fitted to LGVs

- The service brake
- The secondary brake
- The parking brake

The service brake is the principal braking system used and is operated by a foot control. It's used to control the speed of the vehicle and to bring it safely to a halt. It may also incorporate an anti-lock braking system (ABS).

The secondary brake may be combined with the footbrake or the parking brake control. This brake is provided for use in the event of the service brake failing. The secondary brake normally operates on fewer wheels than the service brake and therefore has a reduced performance level.

The parking brake is usually a hand control and may also be the secondary brake. It should normally only be used when the vehicle is stationary. The parking brake must always be set when the driver leaves the driving position. It's an offence to leave any vehicle unattended without applying the parking brake.

LGVs are also frequently equipped with endurance braking systems (commonly called 'retarders').

Anti-lock braking systems

Anti-lock braking systems employ wheel speed sensors to anticipate when a wheel is about to lock under braking. Just before this would happen the system releases the brake and then rapidly re-applies it. This action may happen many times a second to maintain braking performance.

Preventing the wheels from locking means that you can continue to steer the vehicle during braking. However, an ABS is only a driver aid. It doesn't remove the need for good driving practices such as anticipating events and assessing road and weather conditions. You still need to plan well ahead and brake smoothly and progressively.

Anti-lock braking systems are in common use on LGVs and are required by law on some. You'll need to know which vehicle combinations are required to have an ABS fitted by law. Care needs to be taken to ensure that the braking system on the tractive unit or rigid towing vehicle is compatible with the braking system on the semi-trailer or trailer.

Checking ABS It's important to ensure that the ABS is functioning **before** setting off on a journey. Driving with a defective ABS may constitute an offence.

Modern anti-lock braking systems require electrical power for their operation. Multi-pin connectors are required to carry the electrical supply to operate the trailer brakes. The satisfactory operation of the ABS can be checked from the warning signal on the dashboard. A separate signal for the trailer is provided on the dashboard, although in some cases a signal on the trailer headboard will operate instead. The way the warning lamp operates varies between manufacturers, but with all types of signal it should be displayed when the ignition is switched on and should go out no later than when the vehicle has reached a speed of about 10 kph (6 mph).

Endurance braking systems

Commonly referred to as 'retarders', these systems provide a way of controlling a vehicle's speed without using the wheel-mounted brakes. This can be particularly useful when descending long hills as a vehicle's speed can be stabilised without using the service brakes.

Retarders operate by applying resistance, via the transmission, to the rotation of the vehicle's driven wheels. This may be achieved by

- Increased engine braking
- Exhaust braking
- Transmission-mounted electromagnetic or hydraulic devices

Braking generates heat in the brakes and, at high temperatures, braking performance can be affected. The retarder leaves the service brake cool for good performance when required.

The system may be operated in unison with the service braking control (integrated) or by using a separate hand control (independent). Retarders normally have several stages of effectiveness depending on the braking requirement. With independent systems the driver has to select the level of performance required.

When driving on slippery surfaces care must be exercised when operating independent retarders if rear wheel locking is to be avoided. Some retarders are under the management of the ABS system to help avoid this problem.

Connecting a system

It's vitally important that you understand the rules that apply to connecting and disconnecting the brake lines on either an articulated vehicle or a rigid vehicle and trailer combination. You'll be asked to demonstrate this during your practical driving test.

There are two brake configurations that you may encounter – either the three-line or the two-line system. A three-line system comprises

- Emergency **red** line
- Auxiliary **blue** line
- Service **yellow** line

A two-line system has only an emergency and a service line.

Two-line vehicles and two-line trailers are obviously compatible, as are three-line vehicles and three-line trailers. A two-line motor vehicle can be connected to a three-line trailer – the trailer auxiliary line being left unconnected.

When connecting a three-line motor vehicle to a two-line trailer it's important that you follow the vehicle manufacturer's advice as to what to do with the third (blue) line. Failure to follow the manufacturer's instructions could render the combination dangerous.

Safety

Air brake systems are fitted with warning devices that will be activated when air pressure drops below a predetermined level. In some circumstances there may be sufficient pressure to release the parking brake even though the warning is showing. In these cases the service brake may be ineffective. Therefore you should never release the parking brake when the brake pressure warning device is operating.

Towing vehicles are equipped with braking lines for attachment to a trailer. On modern vehicles these lines are fitted with automatic sealing valves rather then manual taps. When a trailer is coupled to a towing vehicle it's important to check that the brakes of the trailer function correctly. If they don't, remedial action must be taken **before** the vehicle is driven. Failure to do so could result in the loss of braking effectiveness on the whole combination.

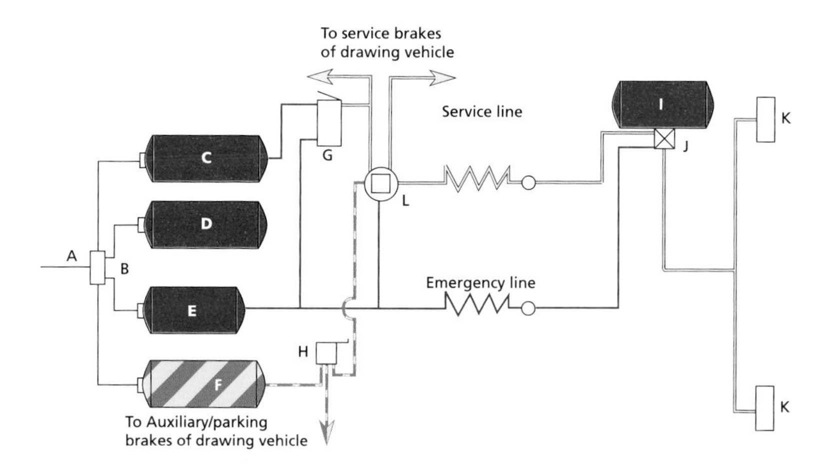

A – Supply from compressor
B – Multi-protection valve
C – Service reservoir (front)
D – Service reservoir (rear)
E – Service reservoir (trailer)
F – Parking/Auxiliary reservoir
G – Dual foot valve
H – Hand control valve
I – Trailer reservoir
J – Relay emergency valve
K – Single diaphragm actuators
L – Trailer control valve (or triple relay valve)

Example of an acceptable two-line connection: two-line vehicle drawing a two-line trailer

Inspection and maintenance

You aren't expected to be a mechanic, however there are braking system checks that *are* your responsibility. If taps (or hand-operated valves) are fitted you must ensure that they're opened after re-connecting and closed before disconnecting brake lines.

When coupling a motor vehicle fitted with automatic sealing valves in the 'Suzie' lines ensure that the trailer is equipped to actuate them. With some spring brake systems the brakes will be applied to the trailer when the emergency line is released. On others the brakes will come off as air seeps away.

Before disconnecting any brake line ensure that the trailer parking brake has been correctly applied. This precaution **must not** be overlooked. There have been a number of fatal accidents due to trailer brakes 'lifting off' as the brake line was released, thus having allowed the trailer to move.

Air reservoirs Air braking systems draw their air from the atmosphere, which contains moisture. This moisture condenses in the air reservoirs and can be transmitted around a vehicle's braking system. In cold weather this can lead to ice forming in valves and pipes and may result in air pressure loss and/or system failure. Some air systems have automatic drain valves to remove this moisture, while others require daily manual draining. You should establish whether your vehicle's system reservoirs require manual draining and, if so, whose responsibility it is to make sure that it's done.

Controls Before each journey make sure that all warning systems are working. Brake pressure warning signals may be activated automatically when the ignition is turned on (as for an ABS) or may require that you use a 'check' switch provided on the driving controls.

Never start a journey with a defective warning device or when the warning is showing. If the warning operates whilst you're travelling, stop as soon as you can do so safely and seek expert help. Driving with a warning device operating may be very dangerous and is an offence.

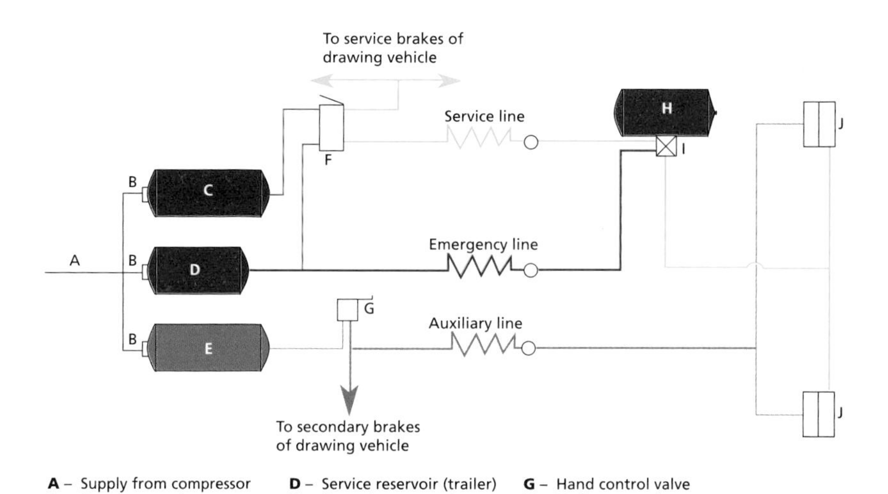

Example of an acceptable brake line connection: three-line vehicle drawing a three-line trailer

Securing a load

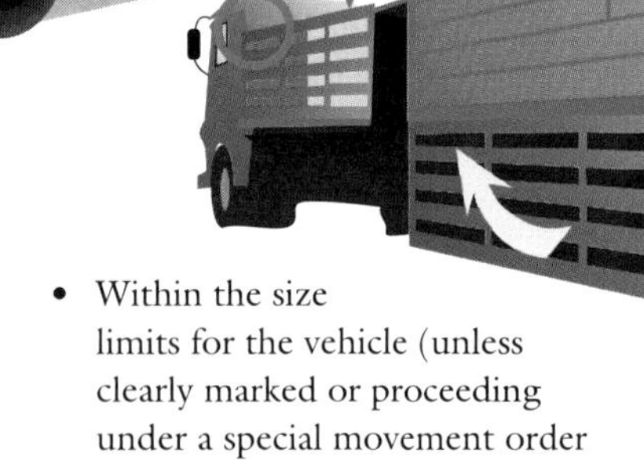

When securing a load you need to take into account

- The nature of the load
- The suitability of the vehicle
- The stability of the load
- The type of restraint
- Protection from weather
- Prevention of theft
- Ease of delivery

The object is to ensure a secure load and a stable vehicle when

- Braking
- Steering

even in emergency situations.

The failure of tyres on the vehicle or trailer shouldn't cause the load to become insecure. This is particularly important when stowing loads such as wooden pallets, hay, etc., which are usually stacked high on flat-bed vehicles.

Any load must be carried so that it doesn't endanger other road users at any time. It should be

- Securely stowed
- Within the weight limits permitted for your vehicle
- Within the size limits for the vehicle (unless clearly marked or proceeding under a special movement order under escort)

You should ensure that all devices for securing the load are effective

- Ropes, chains, straps are secure
- Sheets fastened down
- Container locking handles secured
- Doors, drop sides, tailgates are fastened
- Hatches on tank vehicles are closed to prevent spillage

You should also prevent

- Material falling from bulk cement vehicles
- Losing any nets covering skip loads

Types of load

A load may consist of large heavy pieces of machinery but that doesn't mean that it will stay in place throughout a journey. Fatal accidents have occurred through such items falling from a vehicle or shifting under braking or cornering.

When making a decision about the type of restraints to be used consider what might happen if you have to brake hard and swerve to avoid an accident. Your vehicle might have to negotiate

- Road works
- A construction site
- A lorry park

where an uneven surface may cause it to tilt over.

Material packed in plastic sacks and loaded onto pallets may be liable to slip unless 'shrink-wrapped' or secured by banding. However, material in canvas sacks may well remain totally stable.

Vehicles being carried piggy-back must always have some form of chocks applied to their wheels, in addition to a restraint. Never rely on merely a handbrake holding them in place.

Tubular loads such as scaffolding poles, lamp standards, extrusions, girders, etc. may all move forward with some force if emergency braking occurs. In such cases the headboard on the vehicle or semi-trailer can be demolished, with fatal results.

Dual-purpose trailers have been developed that incorporate a

- Belly tank installed along the centre of the trailer for transporting fluids
- Flat-bed deck above the tank for the conventional carriage of goods

In such instances care must be taken not to rupture the tank below.

Types of restraint

It's important that the correct anchoring points are employed irrespective of the type of restraint being used. Remember, however, that the hooks fitted under some decks are only intended for fastening sheeting ropes.

Ropes

Traditionally, ropes have been the commonest method of securing both a load and sheets. Ropes may be of fibre or modern man-made materials such as nylon, polypropylene, etc.

Whatever type of rope is used you should gain experience in the correct methods of securing the load. The knots used are known in the trade as 'dolly knots'. These can only be released when required (and not otherwise). Additionally, you should ensure that the proper tension is applied and that only the correct securing points are used.

Ropes are totally unsuitable for some loads, such as steel plates, scrap metal, etc.

Straps

These are generally made of webbing and are frequently used to secure many types of load.

Ensure that all straps, tensioners, etc. are kept in good, serviceable condition. If a load has sharp edges, straps with suitable sleeves and corner protectors can be used.

Battens and chocks

Large, heavy objects such as metal ingots, castings, fabrications, etc. should be chocked by nailing battens to the vehicle or trailer deck.

Chains

If there's any danger of either the weight of the load being too great for ropes or straps, or the load having sharp edges that would shear ropes or straps, then chains must be used together with compatible tensioning devices.

Chains will provide added security when tree trunks or logs are being carried. Don't rely solely on vertical stanchions to hold the load.

Sheeting

If sheeting is used – whether tarpaulin, plastic, nylon or any other material – it must be secured in such a way that it doesn't become loose and create a hazard to other road users.

When starting to cover a load with more than one sheet it's sensible to start with the rear-most sheet first, working forward. This type of overlap will reduce the possibility of wind or rain being forced under the sheeting as the vehicle travels along in bad weather conditions.

In order to secure the sheets onto a load you'll need to use the same type of knots used when restraining loads (dolly knots). These remain taut in transit but can be released with the minimum of effort.

All spare sheets and ropes must be tied down securely so that they don't fall into the path of following traffic.

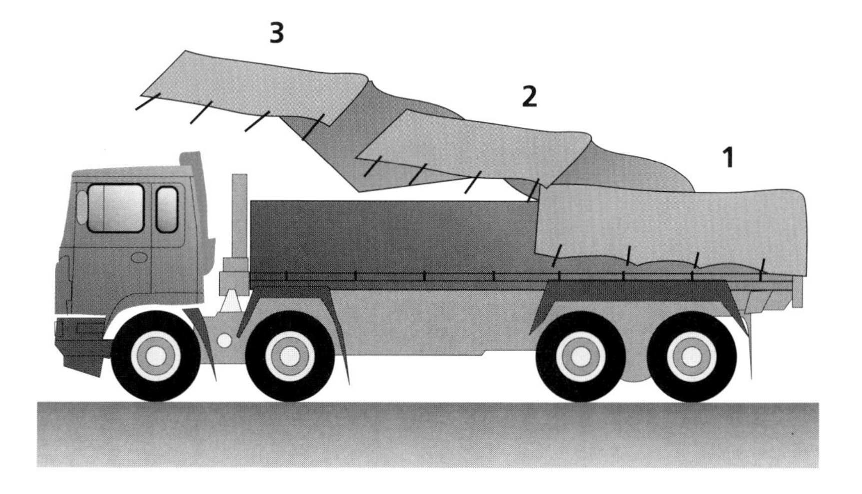

Curtain sides

The manufacturers of vehicles fitted with curtain-side bodies may be satisfied that a high degree of protection is given by the material used in their construction. This, however, doesn't relieve the driver of the responsibility for ensuring that a load is properly stowed and secured so that it won't move while in transit. This is particularly important where there may be a 'multi-drop' load of varying materials, some of which may come under the hazardous materials classification.

Take notice of warnings of poor weather conditions broadcast on the radio, especially if your vehicle is empty. Under such conditions it's often safer to secure both curtain sides at one end of the vehicle, cutting down the wind resistance and removing the likelihood of being blown over or off the road.

Container lorries

ISO (International Standards Organisation) cargo containers should only be carried on vehicles or trailers equipped with the appropriate securing points, which are designed to lock into the container body. Such vehicles may be intended for carrying

- A single 12 metre (40 feet) container
- One or two 6 metre (20 feet) containers
- Larger numbers of smaller, specially designed units

Whatever type of container is carried all locking levers must be in the secured position during transit.

Steel ISO containers shouldn't be carried on flat-bed platform vehicles where there are no means of locking the container in position. Never rely on the weight of the container and its contents to hold it in place on a flat deck.

Ropes are totally inadequate to hold a typical seagoing steel container in place. Skeletal vehicles or trailers that have a main chassis frame with outrigger supports, into which the ISO container can be locked, are safer and more secure.

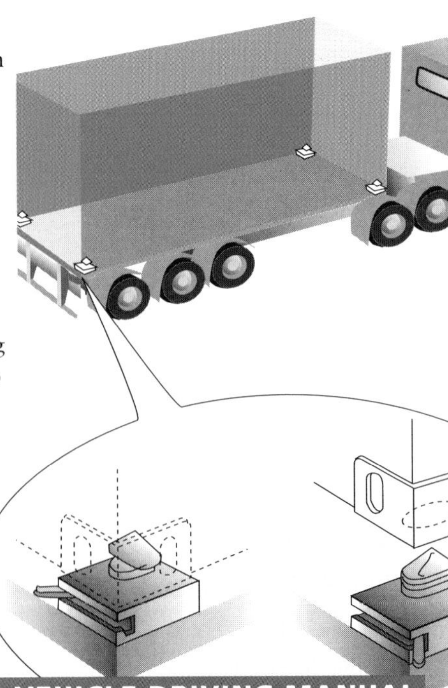

Road-friendly suspension

Reference has already been made to the requirement that some form of road-friendly suspension be fitted to vehicles that are intended to carry increased weights. By replacing springs with some form of compressible material (usually air) a reduction in vibration caused by the impact of LGV wheels on road surfaces will reduce the damage to

- The road surface itself
- Adjacent structures
- Under-road services (gas, water, etc.)
- Bridges

These systems can in some instances be fitted retrospectively to vehicles. However, this does mean that the vehicle will usually have to be equipped with additional compressed air storage tanks, creating some additional weight.

An increasing number of manufacturers are making use of the benefits that road-friendly suspension gives in reducing damage to goods in transit.

Specialised semi-trailers used to carry fragile goods have no rear axles as such. When loading takes place

- The hollow trailer body is positioned to surround the racks holding a fragile load (e.g., glass)
- The body is lowered into place
- The load is secured
- The body is raised into the travelling position

The whole process is carried out by controlling the sophisticated road-friendly suspension system on each trailer wheel assembly.

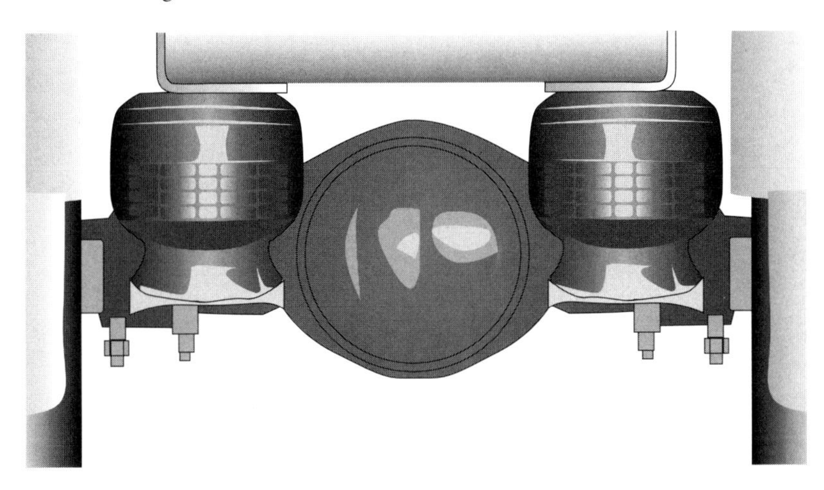

Fuel consumption

Fuel consumption can depend on the design of your vehicle.

- Cab-mounted wind deflectors can effectively lower wind resistance created by large box bodies, together with lower side-panel 'skirts'
- Tipper bodies with prominent strengthening ribs on the outside can be plated-over to give improved performance
- A fly sheet tightly fastened over the top of tipper bodies (especially when empty) can reduce the 'drag' effect

Further information can be found in the Department of Energy booklet *Fuel Efficiency 20: Energy Efficiency in Road Transport.*

You have a part to play in helping to reduce the impact road transport has on the environment.

- Plan routes to avoid busy times and congestion
- Anticipate well ahead
- Avoid over-accelerating
- Brake in good time
- Reduce your overall speed
- Avoid the need to 'make up time'
- Cover bulky loads with sheets to reduce wind resistance

You should also ensure that maintenance schedules are strictly followed

- Filters are changed regularly
- Exhaust emissions meet current regulations
- Diesel injectors are operating efficiently
- Brakes are correctly adjusted
- Tyre pressures are correct

Members of the public are encouraged to report any vehicle emitting excessive exhaust fumes.

Diesel spillages

Because of the extremely slippery characteristics of diesel fuel, care must be taken at all times to avoid spillages. Not only is diesel fuel dangerous to anyone stepping onto it (especially getting down from a vehicle cab), but it also creates a serious risk to other road users, especially motorcyclists.

Take care when refuelling and ensure that all filler caps and tank hatches are properly closed and secure.

Alternative fuels

Diesel has traditionally been the fuel for LGV engines. However, the possibility of using a number of alternative fuels is being explored.

Compressed Natural Gas (CNG)

While there are improvements in the quality of exhaust emissions produced some of the technical disadvantages relate to the size and design of the fuel tanks required.

Methane

Because of the naturally occurring sources of this fuel it's also being considered as a possible alternative to diesel oil, which is a finite resource.

Hydrogen

This is another possible fuel source for road vehicles that's being studied. Again, technical problems include the storage of this highly inflammable gas.

Hazardous goods

Strict regulations regarding the stowage and labelling of materials classified as 'hazardous' must be observed. Under the Road Traffic (Training of Drivers of Vehicles Carrying Dangerous Goods) Regulations 1992 known as the Driver Training Regulations (DTR), the operator of any vehicle used to carry dangerous goods must ensure that the driver of any such vehicle receives adequate instruction and training* to enable them to understand

- Any emergency action that may be needed and the dangers which may be created by the substances being carried
- The driver's duties under the Health and Safety at Work Act 1974
- Any duty imposed by legislation that applies to carrying dangerous goods by road

***Note**: 'Instruction' is the verbal and/or written information that must be given to a driver about their load(s) each day.

'Training' is the formal tuition of drivers, providing them with education about the hazards dangerous substances may present.

The following groups of drivers need to be in possession of a certificate showing that they're licensed by the DVLA to carry dangerous goods by road

- Drivers of road tankers with a capacity of more than 3,000 litres or a maximum permissible weight of over 3.5 tonnes
- Drivers of vehicles carrying tank containers, regardless of the maximum permissible weight of the vehicle
- Drivers of all vehicles carrying explosives (subject to limited exemptions)
- Drivers of all vehicles over 3.5 tonnes maximum permissible weight that are subject to The Road Traffic (Carriage of Dangerous Substances in Packages, etc.) Regulations 1992 [known as PGR]

The DVLA will only issue the certificate upon receipt of proof of attending a course at an approved training establishment and passing the 7357 examination of The City and Guilds of London Institute. The certificate is valid for five years.

Audible warning systems

As an LGV driver it's up to you to recognise the effects your vehicle, and the way in which it's driven, can have on the environment around you.

Reversing your vehicle can cause a hazardous situation. There may be pedestrians in the area that you'll need to warn. There are different types of audible warning device which give a signal to others around the vehicle that it's reversing, such as a

- Bleeper
- Horn
- Recorded verbal message, etc.

These must not be allowed to operate on a road subject to a 30 mph speed limit between 11.30 pm and 7 am.

Take care when setting any vehicle security alarm system. You only want such an alarm to sound when it's necessary – not by mistake.

And remember, using an audible warning device doesn't take away the need to practise good, all-round effective observation. If you think that you're unable to reverse safely you must get someone to help you.

Drivers' hours and records

Goods drivers' hours of work are controlled in the interests of road safety, drivers' working conditions and fair competition. A European regulation sets maximum limits on driving time and the minimum requirements for breaks and rest periods. These are known as the EC rules. Drivers who break the rules are subject to heavy fines and could lose their licence to drive LGVs. Altering drivers' hours records with intent to deceive, or tampering with tachographs can lead to a prison sentence. Similar penalties apply to those who permit such offences.

EC rules

The EC rules apply to vehicles used for the carriage of goods, whether loaded or not, on national and international journeys throughout the European Union. They're also consistent with the rules adopted by many countries beyond. Tachographs must be used under the EC rules.

Exemptions

The following are exempt from EC drivers' hours and tachograph rules. In most of these cases domestic rules apply.

- Vehicles used for the carriage of goods where the maximum permissible gross weight of the vehicle, including any trailer or semi-trailer, doesn't exceed 3.5 tonnes
- Vehicles with a maximum authorised speed not exceeding 30 kph (about 18.6 mph)
- Vehicles used by or under control of the armed services and forces responsible for maintaining public order
- Vehicles used in connection with the sewerage, flood protection, water, gas and electricity services, highway maintenance and control, refuse collection and disposal, telegraph and telephone services, carriage of postal articles, radio and television broadcasting, and the detection of radio or television transmitters and receivers
- Vehicles used in emergencies or rescue operations
- Specialised vehicles used for medical purposes
- Vehicles transporting circus and fun-fair equipment
- Specialised breakdown vehicles
- Vehicles undergoing road tests for technical development, repair or maintenance purposes and new or rebuilt vehicles which haven't yet been put into service
- Vehicles used for non-commercial carriage of goods for personal use
- Vehicles used for milk collection from farms, and for the return to farms of milk containers or milk products intended for animal feed

Drivers are also exempt from the EC drivers' hours and tachograph rules when engaged in the following operations in the UK. In most of these cases domestic drivers' hours rules apply.

- Vehicles used by agricultural, horticultural, forestry or fishery concerns undertaking the carriage of goods within a 50 km radius of the place where the vehicle is normally based. (In the case of fishery undertakings, the exemption applies only to the movement of fish from landing to first processing on land, and of live fish between fish farms.)
- Vehicles used for carrying animal waste or carcasses that aren't intended for human consumption
- Vehicles used for carrying live animals from farms to local markets and vice versa, or from markets to local slaughterhouses
- Vehicles used and specially fitted for such uses as shops at local markets or for door-to-door selling; for mobile banking, exchange or saving transactions; for worship; for the lending of books, records or cassettes; for cultural events or exhibitions
- Vehicles with a permissible weight of not more than 7.5 tonnes carrying material or equipment for the driver's use in the course of his or her work within a 50 km radius of the place where the vehicle is normally based, provided that driving the vehicle doesn't constitute the driver's main activity
- Vehicles operating exclusively on islands not exceeding 2,300 sq km in area, which aren't linked to the rest of Great Britain by a bridge, ford or tunnel open for use by motor vehicles
- Vehicles with a maximum permissible weight of 7.5 tonnes (including batteries) used for the carriage of goods and propelled by means of gas or electricity
- Vehicles used for driving instruction with a view to obtaining a driving licence but excluding instruction on a journey connected with the carriage of a commercial load
- Vehicles operated by the Royal National Lifeboat Institution
- Vehicles manufactured before 1 January 1947
- Vehicles propelled by steam
- Vehicles used by health authorities as ambulances or to carry staff, patients, medical supplies or equipment
- Vehicles used by local authority Social Service departments to provide services for the elderly or the physically or mentally handicapped
- Vehicles used by HM Coastguard and lighthouse services

- Vehicles used by harbour or airport authorities if the vehicles remain wholly within the confines of ports or airports
- Vehicles used by British Waterways Board when engaged in maintaining navigable waterways
- Tractors used exclusively for agricultural and forestry work

Tachographs

When driving within the EC rules, drivers' hours and rest periods are recorded by means of a chart that's inserted into a tachograph. A tachograph is a machine that records the time you spend on tasks during your working day. It can also record the speed that the vehicle travels.

The tachograph should be properly calibrated and sealed by an approved vehicle manufacturer or calibration centre. These must be checked at a Department of Transport (DOT) approved calibration centre every two years and recalibrated every six years. A plaque either on or near the tachograph will say when the checks were last carried out.

If there's anything wrong with the tachograph it should be replaced or repaired by a DOT-approved centre as soon as possible. If the vehicle can't return to base within a week of failure of the tachograph or of the discovery of its defective operation, the repair must be carried out during the journey. While it's broken you must keep a manual record on the chart.

Charts You must carry enough charts with you for the whole of your journey. You'll need one for every 24 hours. You should also carry some spares with you in case the charts become dirty or damaged. Your employer is responsible for giving you enough clean charts for the tachograph installed in the vehicle.

You, the driver, must ensure that the correct information is recorded on the charts. Before departing on your journey you must record

- Your surname and first name
- The date and the place where use of the chart begins (before departing) and ends (after arrival)
- The registration number of the vehicles driven during the use of the chart. (This should be entered before departing in a different vehicle.)
- The odometer reading at the start of the first journey and at the end of the last journey on the chart (and the readings at the time of any change of vehicle)
- The time of any change of vehicle

Recording information The tachograph will start recording onto the chart as soon as it's inserted. You should ensure that the mode switch is in the appropriate position. The modes are shown as symbols.

- Driving symbol

- Other work symbol

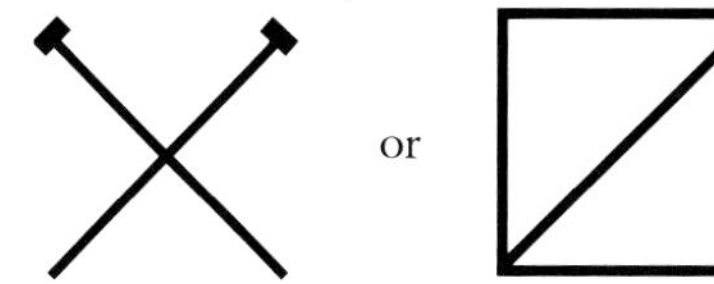

- On duty and available for work

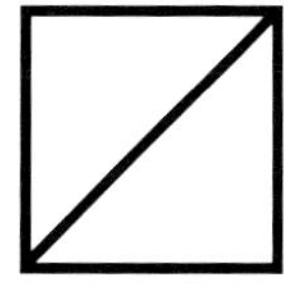

- Break or rest symbol

If you're driving more than one vehicle in one day you must take your chart with you and use it in the next vehicle. If for some reason the equipment in the other vehicle isn't compatible you should use another chart. If your vehicle is likely to be used by someone else while you're away from it you should take the chart with you and manually record on the reverse of the chart other work or rest.

If your chart is dirty or damaged you should start another and then attach it to the damaged one.

Ensure that all the information for the day is entered on your chart(s).

The obligation to record all the complete information correctly falls on you, the driver, as well as on the operator. There are heavy fines imposed for the misuse or falsification of charts.

Chart inspections Your tachograph records must be available for inspection by the enforcement authorities. You must carry your completed charts for the current week and the last day of the previous week on which you drove.

If your records are kept by an enforcement officer you should ask the officer to endorse the replacement chart with the officer's

- Name
- Telephone number

It should also state the number of charts retained. Alternatively, the officer may provide you with a receipt.

To ensure that all records are kept up to date and available for inspection by enforcement staff you must give the completed charts to your employer within 21 days.

EC drivers' hours

'Driving' means being at the controls of a vehicle for the purposes of controlling its movement, whether it's stationary or moving.

Daily driving A day is defined as any period of 24 hours beginning when you start work or driving after the last daily or weekly rest period. The maximum daily hours you may drive is nine. This can be increased to ten hours twice a week. The basic nine hours must be between

- Two daily rest periods or
- A daily rest period and a weekly rest period

Any driving off the public roads doesn't count as driving time. In this case you should record the time as 'other work'.

You must ensure that you take a break of 45 minutes after four and a half hours of driving. This break can be replaced by two or three breaks of no less than 15 minutes during or after the driving period. The total of these shorter breaks must add up to at least 45 minutes in 4.5 hours. During any break you must not drive or undertake any other work.

Daily rest periods A rest period is a time of at least one hour when you're free to dispose of your time as you wish.

You must have a minimum of 11 consecutive hours' daily rest. This can be reduced to nine hours, but not more often than three days a week. In this case you must take the equivalent rest before the end of the following week.

Daily rest can also be taken as 12 hours in two or three periods. In this case each rest period must be at least an hour and the last period must be at least eight hours.

If you're taking your rest period on a ferry or train the daily rest period may be interrupted for up to one hour, but only once. If it is, two hours must be added to the rest time. One part must be taken on land, either before or after the journey. The other part can be taken on board a boat or train. In this case you must have access to a bunk or couchette for both rest periods.

Weekly driving A week is defined as a period between 0.00 on Monday and 24.00 on the following Saturday.

A weekly rest period must be taken after no more than six daily driving periods. You can drive up to 56 hours between weekly rest periods. You must not, however, drive over 90 hours in any one fortnight.

Weekly rest periods When taking the weekly rest period, a daily rest period must normally be extended so that you get at least 45 consecutive hours of rest. You can reduce this to a minimum of 36 hours if you take the rest either where the vehicle is normally based or where you're based. If it's taken elsewhere it can be reduced to a minimum of 24 consecutive hours.

If you take reduced rest you must make up for it by an equal period of rest added to a weekly or daily rest period. This must be taken in one continuous period before the end of the third week following the week in question.

A weekly rest period that begins in one week and continues into the following week may be added to either of these weeks.

Catching up on reduced rest If you've reduced your daily and/or weekly rest periods the compensatory rest must be added to another rest of at least eight hours. You can request to take this at either your base or where your vehicle is based. Rest taken as compensation for the reduction of a weekly rest period must be taken in one continuous block. This compensatory rest period can be made up of any combination of breaks of at least one hour.

Two or more drivers During each period of 30 hours, each driver must have a rest period of not less than eight consecutive hours. There must always be two or more drivers travelling with the vehicle for this rule to apply. A driver may take a break while another driver is driving, but not a daily rest period.

In the interest of road safety all rules regarding drivers' hours should always be followed. However, there might be an emergency situation where you have to depart from the drivers' rules to ensure the safety of people, the load or the vehicle. In these unusual situations you should note the reasons on the back of the tachograph chart.

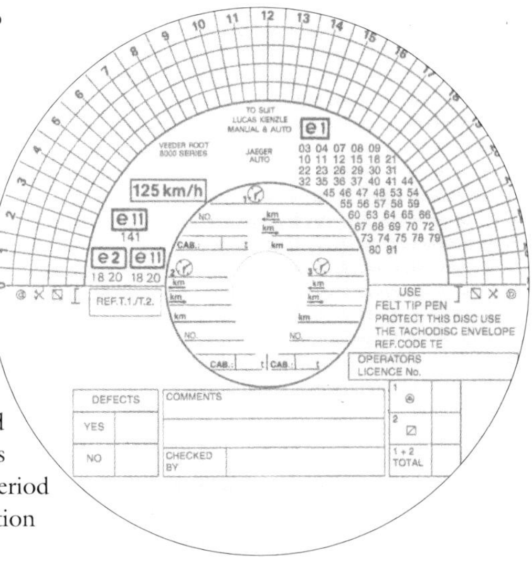

Domestic drivers' hours

The domestic rules apply to most goods vehicles that are exempt from EC rules. There are driving limits, duty limits and road requirements.

Driving limits

You must not drive for more than ten hours in any one day. This limit applies to the time actually spent driving. Off-road driving counts as duty rather than driving.

Daily duty

You must not be on duty for more than 11 hours on any working day. You're exempt from the daily duty limit if you don't drive on any working day or if you don't drive for more than four hours on each day of the week.

Exemptions

- Drivers of vehicles used by the armed forces, the police and fire brigades
- Drivers who always drive off the public roads
- Private driving

Drivers of dual-purpose vehicles or those not exceeding 3.5 tonnes permissible weight are exempt from the duty limit, but not the driving limit.

Vehicles driven by

- Doctors
- Dentists
- Nurses
- Midwives
- Vets

are also exempt.

Vehicles used for any inspection, cleaning, maintenance, repair, installation or fitting

- By a commercial traveller
- By the AA, RAC or RSAC
- For cinematograph or radio and television broadcasting

are included among the exemptions.

The domestic rules also allow for events needing immediate action to avoid danger to life or health of people or animals. These rules also allow for the prevention of serious disruption to essential services or for danger to property.

Keeping records

You must keep a written record of your hours of work on a weekly record sheet, which is available from commercial printers. If you're driving a vehicle in excess of 3.5 tonnes permissible maximum weight that carries parcels or postal items you must use a tachograph.

Mixed EC and domestic driving

It's possible that you may drive under EC rules and domestic rules during a week or even one day. You can choose to drive under EC rules for the whole of the time. If you use a combination of the rules you must ensure that both rules are obeyed at the appropriate times.

You can't use the time driving under EC rules as off-duty time. Similarly, you can't claim driving under domestic rules as rest time for EC rules. Remember, any EC driving in a week means that you must take daily and weekly rest periods.

Know the regulations

In addition to the rules and regulations that apply to drivers' hours or vehicles you should be sure that you comply with any regulations which affect your

- Health
- Conduct
- Vehicle
- Driving
- Licence
- Safety

It's essential that you know and keep up to date with the regulations and the latest official advice.

Your health and conduct

Health

Even apparently simple illnesses can affect your reactions. You should be on your guard against the effects of

- Flu symptoms
- Hay fever
- A common cold
- Tiredness

Falling asleep

Incidents where vehicles have

- Left the road
- Collided with broken-down vehicles, police patrols and other persons on the hard shoulder of motorways

have been attributed to falling asleep at the wheel. The introduction of

- Air-suspension drivers' seats
- 'Floating' cab suspension
- Air suspension on vehicles
- Quieter, smoother diesel engines
- More widely adopted sound-proofing materials

have produced a comfortable 'cocoon' environment where you'll spend most of your working day. This can easily cause tiredness.

Be on your guard against boredom on comparatively empty roads or motorways, especially at night. Always

- Take planned rest stops
- Keep a plentiful supply of fresh air circulating in the cab
- Avoid allowing the cab to become unduly warm
- Avoid driving if you aren't 100% fit to drive
- Avoid driving after a heavy meal

Stop at the next lay-by or pull off the motorway (or slip road) if you start to feel tired.

Drugs

Drug abuse has now reached the point where well-known multinational companies have introduced random drug-testing for their drivers. Those drivers who fail such tests may face instant dismissal.

It should be obvious that you must not take any of the drugs that are generally accepted as 'banned substances' whilst driving. These include

- Amphetamines (e.g., 'diet pills')
- Methylamphetamines (MDMA)
- Benzodiazapine (tranquillizers)
- Methaqualone (sleeping pills)
- Barbiturates (sleeping pills)
- Propoxyphane
- Phencyclidine ('Angel Dust')
- Cannabis
- Cocaine
- Heroin
- Morphine/codeine

Unlike alcohol (the effects of which last for about 24 hours) many of the effects of drugs will remain in the system for up to 72 hours.

'Off-the-shelf' remedies Even everyday cold or flu remedies can cause drowsiness. Read the labels carefully. If in doubt, consult either your doctor or pharmacist. If still in doubt, **don't drive**.

Alcohol

It's an offence to drive with more than

- A breath alcohol level in excess of 35 microgrammes per 100 ml
- A blood alcohol level in excess of 80 mg per 100 ml

Don't drink if you're going to drive.

Be aware that alcohol may remain in the body's system for around 24 hours. The effects on your reactions will be evident the next morning and you could fail a breath test.

If you're convicted of such a drink-driving offence while driving an ordinary motor vehicle a driving ban will result in you losing your LGV entitlement and livelihood.

Your vehicle

You must ensure that the vehicle you drive complies with all regulations that relate to it being legally roadworthy. The items covered include

- Brakes
- Lights
- Tyres
- Windscreen wipers and washers
- Horn
- Mirrors
- Speedometer
- Tachograph
- Number plates
- Reflectors and reflective plates
- Exhaust system
- Any coupling gear
- Speed limiter
- Being correctly plated
- Current test certificate (if required)
- Being properly licensed with the appropriate valid disc(s) displayed
- Insurance
- Seat belts*
- Construction and use
- Any load being carried

* Where seat belts are fitted they must be worn.

'Red' diesel fuel is restricted to use for authorised purposes only. Any driver whose vehicle is found to be illegally operating on this fuel will face severe penalties for attempting to evade excise duty. Roadside checks are frequently carried out by HM Customs and Excise officers.

Speed limiters

UK legislation LGVs must be fitted with a speed limiter if they're

- Over 7.5 tonnes MGW and first registered on or after 1 August 1992
- An articulated tractor over 16 tonnes MGW and rigid vehicles over 16 tonnes MGW where the gross train weight is more than 5 tonnes greater than the MGW, and which is
  - first registered on or after January 1988
  - capable of exceeding 60 mph (96.5 kph) without a speed limiter

Speed limiters must be set so that the vehicle can't exceed a maximum powered speed of 60 mph (96.5 kph).

EC legislation LGVs must be fitted with a speed limiter if they're

- Over 12 tonnes and first registered on or after 1 January 1988 and capable of exceeding 56 mph (85 kph) without a speed limiter fitted

Speed limiters must be set at 56 mph (85 kph).

Exemptions Speed limiter requirements don't apply to a vehicle that's

- Being taken to a place where a speed limiter is to be installed, calibrated, repaired or replaced
- Completing a journey in the course of which the speed limiter has accidentally ceased to function
- Used for police, fire or ambulance purposes
- Used for naval, military or air force purposes when used by the Crown or owned by the MOD
- Being used no more than six miles on a public road in any calendar week between land occupied by the vehicle keeper

Your driving

You must drive at all times within the law and comply with

- Speed limits
- Weight limits
- Loading/unloading restrictions
- Waiting restrictions
- Stopping restrictions (clearways)
- Lighting regulations
- Restrictions of access to
  – pedestrian precincts
  – residential areas
  – traffic calming zones
  – play streets
- All traffic signs
- Road markings
- Traffic signals at
  – junctions
  – level crossings
  – fire or ambulance stations
  – lifting or swing bridges
- Signals given by authorised persons
  – police officers
  – traffic wardens
  – local authority parking attendants
  – school crossing patrols
  – persons engaged in road repairs
- Motorway regulations
- Regulations governing specific locations
  – tunnels
  – bridges
  – ferries
- Pedestrian crossing rules

Driving licences

The LGV driving licence is a necessity if you wish to earn your living driving large goods vehicles. It's essential that when you drive any vehicle other than an LGV your driving continues to be up to the highest standards. If you accumulate penalty points on your category B licence eventually your LGV licence will be at risk.

Speeding offences

Police forces and local authorities are now using the most up-to-date technology in an effort to persuade drivers to comply with speed limits.

At some locations fixed cameras that photograph vehicles exceeding the speed limit have been installed. Improved detection equipment can now also 'lock on' to individual vehicles in busy traffic flows. In addition, new electronic systems now display the registration number and speed of any offending vehicle at selected motorway locations with a view to 'showing up' the driver concerned.

Drivers whose speed is considerably higher than the legal speed limit can expect a proportionately higher penalty if a successful prosecution results. But remember, the aim is to improve driving standards, not to increase prosecutions.

Red light cameras

Cameras have been installed at many notorious accident spots to record drivers not complying with the traffic signals. These are also intended to act as a deterrent and to improve safety for road users in general.

Whether it relates to an alleged speeding or traffic signal offence, any photograph produced as evidence and that shows the

- Time
- Date
- Speed
- Vehicle registration number
- Time a red signal had already been showing

will prove difficult to dispute.

Red Routes

On many roads in London yellow lines are being replaced with red lines. A network of priority (Red) routes for London was approved by Parliament in June 1992 as a means of addressing traffic congestion problems and widespread disregard of parking restrictions in the capital. Work is now under way to introduce Red Route measures on 315 miles of London's most important roads by the year 2000. On these roads new Red Route signs and red markings are being introduced to replace the old yellow-line restrictions.

Yellow-line exemptions **don't** apply on Red Routes. During the day loading is only allowed in marked boxes. Overnight and on Sundays most controls are relaxed to allow unrestricted stopping. It's important to check signs carefully as the hours of operation for Red Routes vary from area to area.

Red Route controls are enforced by Metropolitan Police traffic wardens. There's a fixed fine for illegal stopping on a Red Route, with no discounts for early payment.

The police or traffic wardens are able to provide limited dispensations for the rare occasions when it's necessary to load in a 'no stopping' zone. These will be available from the local police station.

There are five main types of Red Route controls.

Double red lines These ban all stopping 24 hours a day, seven days a week. You aren't allowed to stop for

- Loading
- Dropping off passengers
- Visiting shops

Single red lines These ban all stopping during the daytime, such as 7 am to 7 pm Monday to Saturday. Outside these hours unrestricted stopping is allowed.

Parking boxes These allow vehicles free parking and can also be used for loading. Red boxes allow parking or loading outside rush hours, such as between 10 am and 4 pm, for either 20 minutes or an hour during the day. White boxes allow parking or loading at *any* time, but the length of stay may be restricted to 20 minutes or an hour during the day. At other times, such as 7 pm to 7 am and on Sundays, unrestricted stopping is allowed in either type of parking box.

'Loading' is defined as when a vehicle stops briefly to load or unload bulky or heavy goods. These goods must be heavy or bulky enough so that it isn't easy to carry them any distance and it may involve more than one trip. If this is the case then your vehicle should be parked legally and the goods carried to the premises. Picking up items that are able to be carried, like shopping, doesn't constitute loading.

Loading boxes These mark the areas where only loading is allowed. Red boxes allow loading outside rush hours, such as between 10 am and 4 pm, for up to 20 minutes.

White boxes allow loading at *any* time, but during the day the length of stay is restricted to a maximum of 20 minutes. At other times, such as between 7 pm and 7 am and on Sundays, unrestricted stopping is allowed in either type of loading box.

Clearways These are major roads where there's no need to stop. There won't be red lines but Red Route clearway signs will indicate that stopping isn't allowed at any time.

For more information on Red Routes contact the Traffic Director for London, College House, Great Peter Street, London SW1P 3LN. Tel: 0171 222 4545.

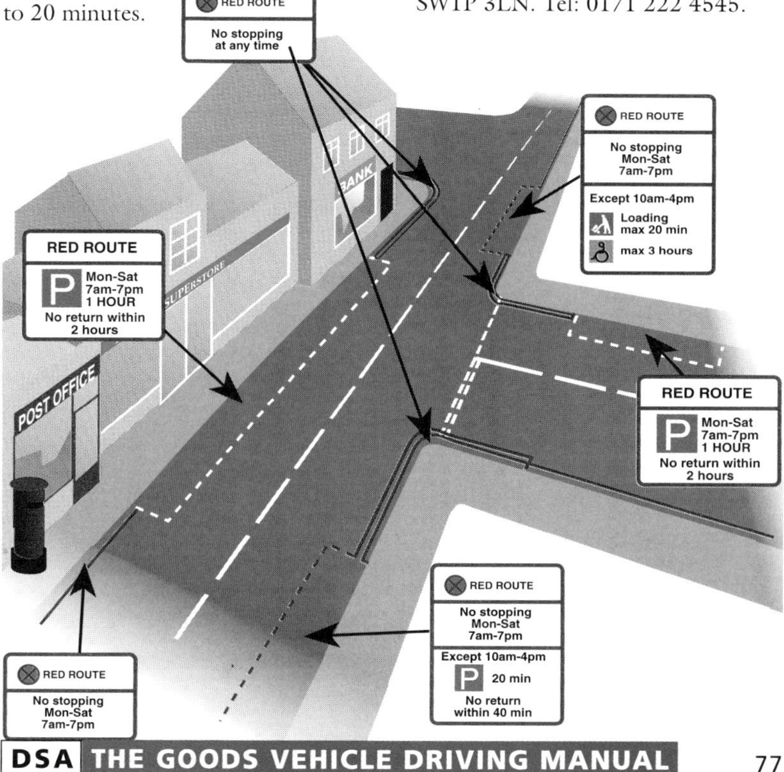

Health and Safety

Many more activities have become the subject of Health and Safety regulations. These include

- Limits to the weight of objects that should be lifted manually, e.g., loading packages
- Provision of protective clothing
  - reflective jackets
  - boots
  - gloves
  - warm clothing
  - hard hats

  where appropriate to the nature of the work

Asbestos

During vehicle maintenance, drivers should be aware of the dangers to health from asbestos dust especially when dealing with components known to contain this material, such as

- Brake shoes
- Clutch plates
- Tank or pipe lagging

Safe working practice

Extra care must be taken when working

- Near or over inspection pits (danger of falling)
- Under hydraulically raised tipper bodies (danger of being crushed – use props!)
- Near engines emitting exhaust fumes (breathing problems)
- With solvents or degreasing agents (lung and skin problems)
- Close to vehicle batteries (risk of burns or explosion)
- At the rear of a vehicle fitted with a 'tail-lift' mechanism (foot injuries)
- In or near paint spray shops (lung problems from vapour)

Anti-theft measures

Instances of theft of vehicles and trailers are unfortunately common. You're responsible for your vehicle, so you should reduce the risk of it being stolen.

- Don't discuss details of your load with any unauthorised person
- Never leave the keys in the cab while it's unattended, even if you're at the rear of the vehicle
- You can't afford to give a lift to anyone, however plausible their story or innocent they look
- Wherever possible, try to avoid using the same route and making the same drops and rest stops
- Have all major components (plus glass) security etched with the vehicle identification number (VIN)
- Only park in secure, well-lit, reputable overnight lorry parks if your rest stops can be planned this way
- One simple but effective measure that many drivers adopt at overnight stops is to park with the rear doors of their vehicle or its trailer/container hard up against another vehicle. This works well on most occasions
- Keep your mobile telephone handset with you, if one is available
- Avoid parking in obviously vulnerable areas if at all possible
- Ensure that all doors are locked and the windows secure if you sleep in the cab overnight
- Always ask to see the identity of any officer who might stop you
- Have an alarm system and/or immobiliser fitted to the vehicle by a reputable security specialist, and approved by the insurance company
- Avoid leaving any trailer unattended unless on approved secure premises
- Fit a kingpin or drawbar lock to any trailer that has to be left unattended

Operators are advised to seek the advice of the local crime prevention officer, especially if engaged in the transit of high-value merchandise.

Recent research has revealed that more than 3,000 LGVs with an insured value of £30 million were stolen in the UK in one year. Most of these vehicles have never been recovered. If you see anything suspicious ring 999 and report it.

PART THREE DRIVING SKILLS

This part looks at professional driving and the skills you need to acquire to help you to deal with various road situations.

The topics covered

- Professional driving
- Driving at night
- Motorway driving
- Bad weather
- Accidents
- Breakdowns

Essential skills

A professional driver should develop the skills necessary to make clear, positive decisions about situations encountered on the road. The skills you'll require are

Control

You should develop the physical skills that enable you to be in control of your vehicle at all times. You should know how your vehicle and its load will handle in any situation you encounter by understanding its capabilities and limitations.

Awareness

You need to know what's happening around you so that you're always conscious of any potential hazards that might develop. This will give you the time to deal with them as they occur.

Planning

Proper planning means that you'll be able to act early when approaching junctions or hazards. This will prevent unnecessary braking and gear-changing, helping you to make progress in traffic. Loaded large vehicles take longer to gain speed than smaller vehicles. Other road users will appreciate your ability to avoid late signalling, constant braking and slow acceleration away from hazards.

Anticipation

By knowing the correct way of dealing with situations as they occur you'll develop anticipation of how to behave in those instances. You'll also have a better insight into the way others respond to those same situations.

It's essential that you're in control of your vehicle at all times. You should drive skilfully and plan ahead so that your vehicle is travelling at the appropriate speed and in the correct position for the next manoeuvre you need to take. You should never have to rush or take action hastily. By adopting the correct techniques you'll create the time and room to complete your intentions safely.

Other road users

Others on the road might make mistakes. You have to accept that other road users aren't always aware of the extra room or time you need, due to the size of your vehicle.

Young children

Young children are particularly unpredictable and might run out into the road suddenly. If you're passing pedestrians who are walking on the pavement but close to the kerb, you must be aware that the size of your vehicle could cause a draught. This could unsteady a small child or, indeed, an adult. Always check your nearside mirror as you pass.

Cyclists

Over a quarter of cyclist fatalities in road traffic accidents result from a collision with an LGV, yet goods vehicles comprise only about 7% of the traffic on UK roads.

You need to allow cyclists as much room as you would a car. They might swerve to avoid a drain cover or a steep camber in the road. If they're approaching a junction or roundabout you must be aware that they might turn right from the left-hand lane, crossing the path of traffic.

The size and shape of your vehicle makes it essential that you're aware of the presence of cyclists **all around** you. Use your nearside mirror as you pass a cyclist to ensure that you've done so safely.

Be aware that when you're waiting at a junction they might move up along either side. If they're positioned in front of your nearside mirror, between the kerb and your front nearside wheel, they'll be difficult to see. You should be aware of this situation as it develops and allow them to move away before you move off.

If you see a cyclist ahead of you glancing round to their right, they're probably about to turn right. Allow for this.

Horses and other animals

Horses are easily frightened by

- Noise
- Vehicles passing too close

If you see horse riders ahead plan your approach carefully. Slow down safely and don't rev the engine. You should allow for the fact that some of the riders might be learners and won't have full control if the animal is startled or frightened. When you pass them do so slowly and leave plenty of room. Always check your nearside mirror to ensure that you've completed the manoeuvre safely.

The elderly

Some elderly pedestrians may have poor eyesight or hearing difficulties. This might make them indecisive and they may sometimes become confused. They also might take longer to cross the road. You need to understand this and allow them more time.

Elderly drivers might be hesitant or become confused at major junctions or gyratory systems. Don't intimidate them by driving up too close or revving the engine.

Learner drivers

Learner drivers who aren't used to all driving situations and other types of road user might be affected by a close-following LGV. They might be driving at an excessively slow speed or be hesitant. Be patient and give them room.

Effective observation

Due to the height of the cab you may have a better view from your driving position than other road users. However, because of its size and design an LGV will have more blind spots than many smaller vehicles.

You should use the mirrors constantly and act upon what you see in them to assess what road users around you are doing or might do next. You must frequently check down the sides of your vehicle.

Offside

- For overtaking traffic coming up behind or already alongside. Do this before signalling
- Before changing lanes, overtaking, turning right or moving to the right

Nearside

- For cyclists or motorcyclists 'filtering' up the nearside
- For traffic on your left when moving in two or more lanes
- To check when you've passed another road user, pedestrians or parked vehicle before moving back to the left
- To verify the position of the rear wheels of the vehicle or trailer in relation to the kerb
- Before changing lanes, after overtaking, turning left or moving closer to the left, leaving roundabouts

You should use your mirrors frequently so that you're constantly aware of what's happening around you.

Because of your relatively high seating position you should also be aware of pedestrians or cyclists who may be directly in front of the vehicle but out of your normal field of vision, especially

- At pedestrian crossings
- In slow-moving congested traffic

Remember, just a simple glance isn't enough. You need to check carefully.

Some LGVs, particularly those with sleeper cabs, give very limited vision to the side. Before moving away wind down the window and lean out and look round to ensure that it's clear before the vehicle starts to move.

Many modern vehicles are fitted with an additional nearside mirror specifically positioned so that the driver can observe the nearside front wheel in relation to the kerb. Use it whenever you're moving off or pulling in to park alongside the kerb, and to check the vehicle's position when you have to move close to the left in normal driving.

Striking the kerb at speed or wandering onto a verge can seriously deflect the steering or damage the tyre.

Mirrors

When you're learning to drive, get into a routine of checking your mirrors. It's important to know as much about traffic conditions all around you as it is about what's going on ahead.

Before you consider changing direction or altering speed you should assess how your actions will affect other road users. Most non-LGV traffic attempting to overtake will normally be catching up to your vehicle at noticeably higher speeds.

You should use the mirrors well before you signal your intention or make any manoeuvre, such as

- Moving away
- Changing direction
- Turning left or right
- Overtaking
- Changing lanes
- Slowing or stopping
- Speeding up
- Opening the cab door

Your mirrors should be

- Clean and free from dust and grime
- Properly adjusted to give a clear view behind. This is particulary important when you're transporting an oversized load that projects over the normal width of the vehicle

Looking isn't enough.

You must act sensibly and positively on what you see. Take note of the speed, behaviour and likely intentions of following traffic.

Take care not to allow your vehicle to 'wander', however slightly, before changing lanes. An LGV occupies much of the available lane width already and any move away from a mid-lane position may cause an overtaking driver or rider to assume that you're starting to pull out into their path.

Blind spots

You might not be able to see much by looking round, especially if the vehicle is fitted with a sleeper cab. This is all the more reason for being continually aware of vehicles just to the rear on either the offside or the nearside in blind-spot positions.

A quick sideways glance is often helpful, especially

- Before changing lanes on a motorway or dual carriageway
- Where traffic is merging from the right or the left
- When approaching the main carriageway from a motorway slip road

Observation at junctions

Despite having a higher seating position than most drivers there will still be some junctions where your view is restricted by parked vehicles.

If it's possible, look through the windows of these vehicles, or if there are shops opposite look for reflections in the windows. If you're still unable to see any oncoming traffic you'll have to ease forward until you can see properly. Do this without encroaching too far into the path of approaching traffic.

Some road users are more difficult to see than others, particularly cyclists – who will generally be approaching close to the kerb from the right. Motorcyclists are often difficult to see and can be travelling fast. Assess the situation. Don't emerge until you know that it's clear.

Pedestrians can often act unpredictably at junctions, running or even just stepping out, oblivious to your presence. Take in the whole scene before you commit yourself to moving a large (and frequently long) vehicle out across the path of oncoming traffic.

If you don't know, don't go.

Zones of vision

As an LGV licence-holder your eyesight must be of a high standard. A skilful driver should be constantly scanning the road ahead and interpreting what's happening or likely to happen.

Always be aware of what's behind and alongside you. Use your peripheral vision to see changes 'out of the corner of your eye' before reacting to them. Look out for the possibility of

- Vehicles about to emerge
- Children running out
- Other pedestrians stepping out

Safe distances

Never drive at such a speed that you can't pull up safely in the distance that you can see to be clear. This should be irrespective of

- Weather
- The road surface
- Any load

Don't drive beyond the limits of your vision.

Keep a safe separation distance between you and the vehicle in front. In reasonable weather conditions leave at least 1 metre (about 3 feet) per mph of your speed, or a two-second time gap. In poor weather, on wet roads, you'll need to at least double the distance and allow a four-second time gap.

Look well ahead

Look well ahead for stop lights. On a road with the national speed limit in force or on the motorway watch for other vehicles' hazard warning lights. These might be flashing to indicate that traffic ahead is slowing down sharply for some reason.

The two-second rule

You can check the time gap by watching the vehicle in front pass an object such as a bridge, pole, sign, etc. and then saying to yourself

'Only a fool breaks the two-second rule.'

You should have finished saying this by the time you reach the same spot. If you haven't finished the rhyme when you pass the spot, you're too close.

On some motorways this rule is drawn to drivers' attention by 'chevrons' painted on the road surface. The instruction 'Keep at least two chevrons from the vehicle ahead' also appears on a sign at these locations.

In congested traffic moving at slower speeds it may not be practicable to leave as much space but you'll still need to leave enough distance in which to pull up safely.

If you find another vehicle driving too close behind you, gradually reduce your speed to increase any gap between you and a vehicle ahead. You'll then be able to brake more gently and remove the likelihood of the close-following vehicle running into the rear of your vehicle.

If another vehicle pulls into the safe separation gap that you're leaving, ease off your speed to extend the gap again.

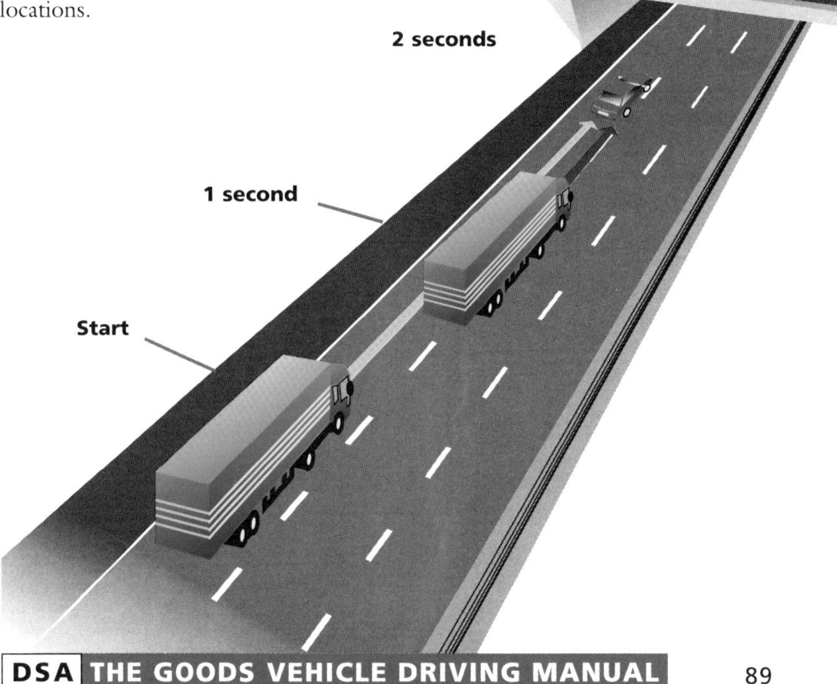

Traffic signals

By planning well ahead you'll ease some of the effort needed to drive an LGV. Anticipating traffic speeds ahead and easing off the throttle means that you may be able to keep your vehicle moving. This will avoid the need to brake, to make a number of gear changes, to come to a stop, or to apply the handbrake. By driving like this you'll be able to make good progress and will keep down fuel costs.

Approaching traffic lights

Signals on green

- Look well ahead and gauge how much traffic is waiting at each side of the junction that you're approaching
- Ask yourself
  - How long has green been showing?
  - If the signals change, am I driving at such a speed that I can stop safely?
  - If I have to brake hard, will following traffic be able to stop safely?
  - Are there any vehicles waiting to turn across my path?
  - How are the road surface and weather conditions going to affect the vehicle's braking distance?

Signals on amber The amber signal means STOP. You may only continue if you

- Have already crossed the stop line
- Are so close to the stop line that to pull up might be unsafe or cause an accident

Signals on red The red traffic signal means that you must stop. You may be able to time your approach so that you're able to keep the vehicle moving as the signals change. This is especially important when driving a laden vehicle uphill to traffic signals. Look well ahead.

Signals not working If you come upon traffic signals that aren't working, or there's a sign to show that they're out of order, treat the location as you would an unmarked junction and proceed with great care.

Don't

- Accelerate to try to 'beat' the signals
- Leave it until the last moment to apply the brakes – harsh braking could result in loss of control

Harsh accelerating or braking could also cause your load to move.

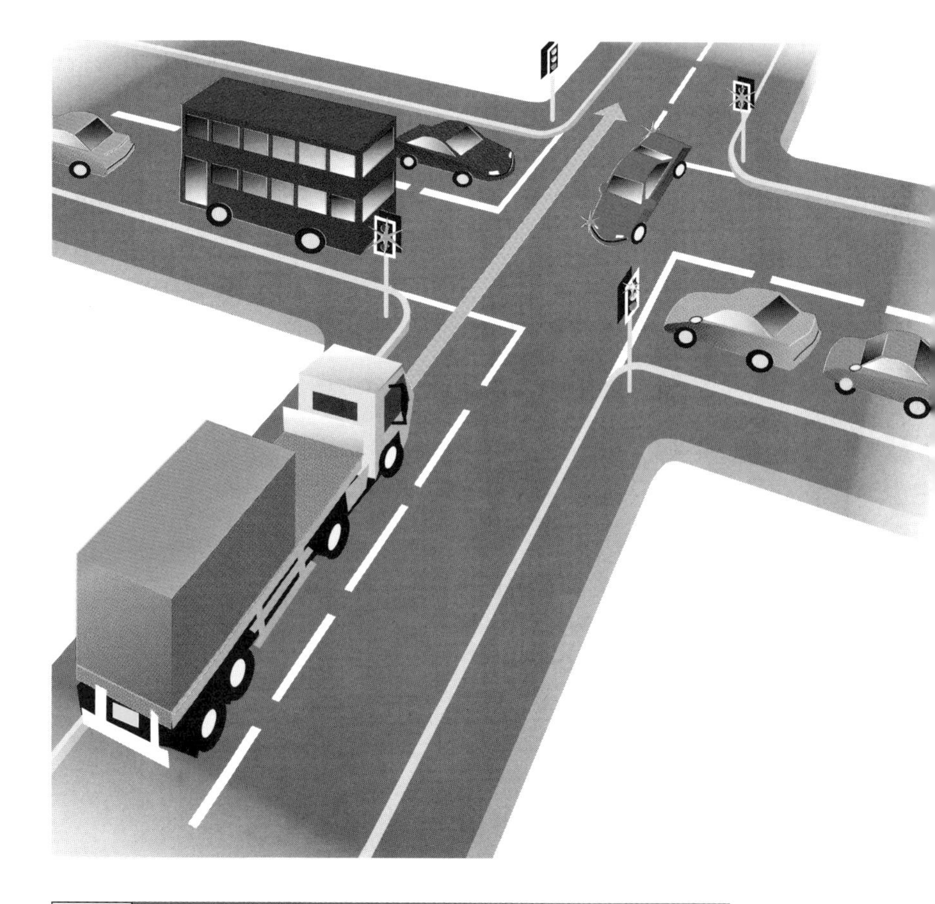

Signalling

You should signal to

- Warn others of your intentions, especially if this involves a manoeuvre not readily apparent to other road users
- Help other road users

Road users include

- Drivers of oncoming vehicles
- Drivers of following vehicles
- Motorcyclists
- Cyclists
- Crossing supervisors
- Police directing traffic
- Pedestrians
- Horse riders
- Road-repair contractors

Give signals

- Clearly and in good time
- That are illustrated in *The Highway Code*

Your signals should be readily understood by all other road users. Try not to mislead others by giving signals that could confuse, especially when intending to pull up just after a road on the left. Another road user might misunderstand the meaning of the signal. In situations like this you should use your common sense and be ready for others' actions.

Don't use the headlights as a signal to give up or take priority. This might lead other vehicles into a hazardous situation. Any signal that doesn't appear in *The Highway Code* is unauthorised and could be open to misinterpretation by another road user.

Using the horn

There are few instances when you'll need to sound the horn. Use it only if you

- Assess that another road user may not be aware of your presence, thus avoiding possible danger
- Need to warn other road users of your presence – at blind bends or a humpback bridge, for example

Sounding the horn doesn't

- Give you priority
- Relieve you of the responsibility to drive safely

Don't use the horn

- When stationary
- At night between 11.30 pm and 7 am in a built-up area, unless there's danger from a moving vehicle
- As a rebuke or simply to attract attention (unless to avoid an accident)

Avoid long, aggressive blasts on the horn, which can alarm pedestrians. In any case, some pedestrians might be deaf.

Problems encountered

Driving an LGV at night, often over long distances, requires additional skills. It also places added responsibilities on the driver.

The problems related to driving at night are

- Much less advance information
- Limited lighting (street lights or vehicle lights only)
- The headlights of oncoming vehicles
- Shadows created by patchy street lighting
- Ineffective lighting on other vehicles, pedal cycles, etc.
- Dangers created by the onset of tiredness

Many deaths have occurred because the driver of a large vehicle either was overcome by tiredness or failed to see an unlit broken-down vehicle until it was too late. Long night journeys, particularly on motorway routes with little to relieve the monotony, require planning and close attention to proper rest and refreshment stops.

Tiredness

Falling asleep at the wheel can happen for only a second or two and yet result in the loss of control.

Be on your guard.

Don't

- Drive without a proper rest period
- Allow the cab to become too warm
- Eat heavy meals just before setting out
- Take your eyes off the road to change radio channels or to change a tape

Try to

- Keep plenty of cool fresh air moving through the cab
- Walk around in the fresh air during a rest stop before setting off again

If you feel your concentration slipping, pull up at the next safe, convenient place.

Night vision

An LGV driver is required to have a better standard of eyesight than other road users. Make sure that your night vision matches up to this higher standard. Have your eyesight checked regularly and avoid

- Wearing tinted glasses at night
- Using windscreen or window tinting sprays

Lighting-up time

You should be prepared to switch on whichever vehicle lights are appropriate to the conditions, regardless of the official lighting-up times. If the weather conditions are poor or it becomes overcast early switch on your lights. See and be seen.

You must drive at an appropriate speed so that you can stop in the distance that you can see to be clear. In most cases that will be within the distance illuminated by your headlights or by street lights.

Unlit vehicles

Vehicles under 1,525kg are permitted to park in 30 mph zones without lights at night time. Be on the alert when driving in built-up areas, especially when the street lighting is patchy.

Builders' skips are required to be lit and show reflective plates to oncoming traffic. Both of these measures can be either neglected or subject to vandalism, so watch out for unlit skips.

Adjusting to darkness

When you step out from a brightly lit area into darkness, such as when leaving a motorway service area, your eyes will take a short while to adjust to the different conditions. Use this time to check and clean your vehicle's lights, reflectors, lenses and mirrors.

At dawn

Other drivers may have been driving through the night and may also be less alert. Leave your lights on until you're satisfied that other road users will see you coming.

It's harder to judge speed and distance correctly in the half-light at dusk and dawn. The colour of some vehicles makes them harder to see in half-light conditions. Switch on your lights to ensure that others can see you.

See and be seen.

Vehicle lighting

It's essential that all lamps are clean and bulbs and light units are operating correctly. In addition to the driver being able to see ahead properly, it's essential that other road users are able to recognise the size of your vehicle and its direction of travel.

All regulation markers and rear lights must be lit and clear of dirt and obstructions such as ropes, sheets, overhanging projections, etc. If you have permission to move any load at night that projects beyond the normal size of the vehicle all additional marker lights and hazards lights should be on.

Oversized loads are usually in lay-bys, etc. overnight. However, in certain circumstances the police authority responsible for that particular area may consider the load would be more safely moved when there's less traffic on the road. Look for any signals given by the escort of such vehicles.

Avoid the 'Christmas tree' effect seen on some vehicles. It can be distracting and confusing to other road users at night. Also, any red light used in the cab must not show to the front of the vehicle.

Auxiliary lighting

LGV drivers must conform to regulations governing the use and fitting of any auxiliary lamps, especially with regard to their mounting height from the road surface. It's an offence to use headlights or spotlights whose centres are less than .6 metres (2 feet) from the ground, except in poor weather conditions such as mist, fog or falling snow.

Any lights showing to the front should be white (or, as allowed on some vehicles, yellow) unless they're side marker lights, required to be fitted by law to certain longer vehicles.

If your vehicle is fitted with any additional working lights to assist coupling/uncoupling, loading, etc. remember to switch them off when the vehicle is out on the road.

High-intensity rear fog lights and additional front fog lights should only be used when visibility is less than 100 metres (about 330 feet).

Amber hazard lights are required, depending on the load projecting beyond specified limits or the vehicle travelling at slower speeds than normal.

Parked vehicles

All LGVs must have lights on when parked on the road at night. A lay-by is usually within the specified distance from any carriageway, so lights are still required.

Unless your vehicle is in an 'off-street' parking location, such as a lorry park, it must be clearly lit to comply with the law.

Driving in built-up areas

Always use dipped headlights in built-up areas at night. It helps others to see you and assists your vision if the street lighting varies or is defective.

Be on the alert for

- Pedestrians in dark clothing
- Runners
- Cyclists (often with poor lighting)

Take extra care when approaching pedestrian crossings. Drive at such a speed that you can stop safely if necessary.

Make sure that you still obey the speed limits even if the roads appear to be empty.

Maintenance work

Essential maintenance work is often carried out at night time. Be on the alert for diversion signs, obstructions and coned-off sections of road when you're driving at night time.

Street cleansing often takes place in the larger cities at night so be on the lookout for slow-moving vehicles.

Driving in rural areas

If there's no oncoming traffic you should use full beam headlights to see as far ahead as possible. Dip your lights as soon as you see oncoming traffic to avoid dazzling the oncoming driver or rider.

If there's no footpath, be on the alert for pedestrians in the road. *The Highway Code* advises pedestrians to walk facing oncoming traffic in these situations, but not all pedestrians follow this advice.

Fog at night

If there's any possibility of fog developing at night **don't drive.**

If the fog becomes so dense that you're unable to go any further safely, your vehicle will present a serious hazard to other vehicles. Because of the difficulties of getting an LGV off the road in dense fog it's better not to start out in the first place.

If you start your journey when there's fog about and you're delayed you'll be committing an offence if you exceed the permitted hours of driving for that period, because the delay was foreseeable.

Overtaking at night

Because LGVs take some considerable time to complete an overtaking manoeuvre you must only attempt one when you can see well ahead that it's safe to do so.

This means that unless you're driving on a motorway or dual carriageway the opportunities to overtake will be limited. Without street lighting you won't be able to assess if there are bends, junctions, hills, etc., which may prevent you seeing an oncoming vehicle.

If you do decide to overtake make sure that you can do so without 'cutting in' on the vehicle being overtaken, or causing oncoming vehicles to brake or swerve.

Never close up on the vehicle ahead prior to attempting to overtake. This will restrict your vision of the road ahead.

Separation distance

Avoid driving so close to the vehicle ahead that your lights dazzle the other driver. Make sure that your lights are on dipped beam.

If a vehicle overtakes you dip your headlights as soon as the vehicle starts to pass you. Your headlights should fall short of the vehicle in front.

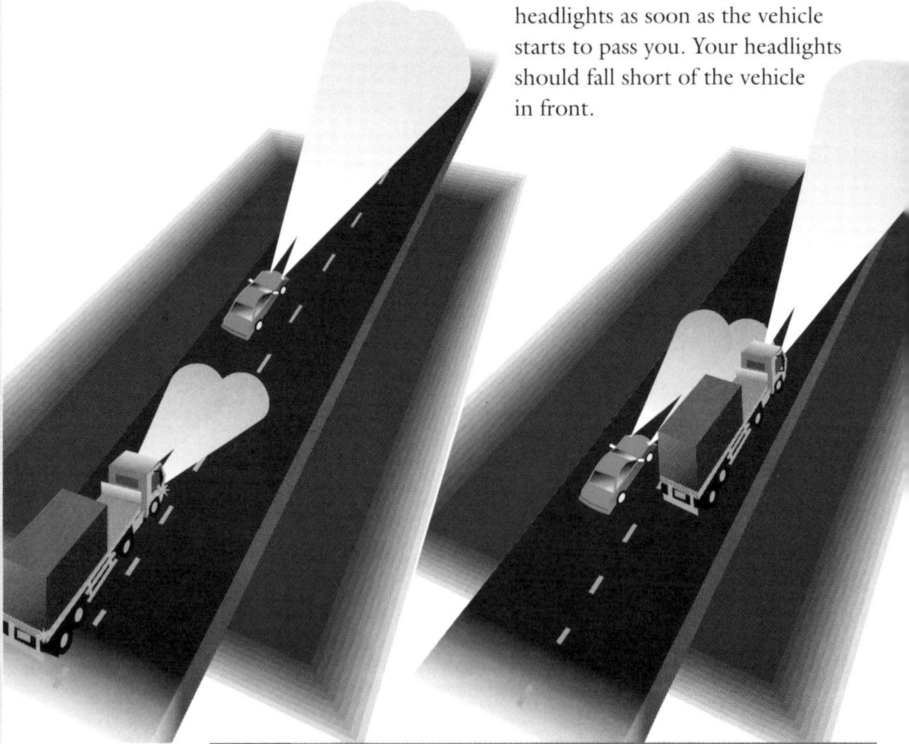

Breakdowns

If your vehicle breaks down try to pull it as far off to the left as possible. If you can get off the main carriageway without causing danger or inconvenience to other road users, especially pedestrians, do so. But don't park on the pavement: the weight of an LGV can damage paving stones and underground services.

If you have a warning triangle place it at least 50 metres (165 feet) behind the vehicle on normal roads or 150 metres (495 feet) on motorways. Some form of warning is vital if an electrical problem has put the rear lights out of action.

Don't attempt to work on the offside of the vehicle unless protected by a recovery vehicle with flashing hazard lights. Even then, take great care on roads carrying fast-moving traffic. Injuries and fatalities have occurred at the scenes of initially simple breakdowns.

If your vehicle is causing an obstruction and possible danger to other road users inform the police as soon as possible. This is particularly important if your vehicle is carrying any hazardous materials.

If you suspect that your vehicle has a mechanical problem don't be tempted to continue on your journey. Small defects could become dangerous if they're left without attention. You could also end up creating traffic chaos if your vehicle eventually breaks down in a difficult location.

Recovery agencies

If you're engaged in long-distance work, especially at night, it's wise to ensure that the vehicle is covered by a reputable recovery agency. The cost of towing or repairing an LGV could be substantial without the benefit of recovery membership.

For safety reasons, vehicles that break down on the motorway are required to be removed as quickly as possible.

Basic preparation

Motorways are statistically the safest road systems in the UK. However, motorway accidents invariably involve a larger number of vehicles travelling at high speeds and usually result in more serious injuries and damage than incidents on normal roads.

Because of the high numbers of LGVs using the motorway network's inter-city links, many of these accidents involve LGVs. But if everyone who used the motorway drove to the same high standard that's required of LGV drivers it's arguable that many of these incidents could be avoided.

The higher overall speeds and the volume of traffic cause conditions to change much more rapidly than on normal roads. For this reason you need to be

- Totally alert
- Physically fit
- Concentrating fully

If you aren't, you may not be able to react to any sudden change taking place ahead of you.

Fitness

Don't drive if you're

- Tired
- Unwell
- Taking flu remedies, etc.
- Emotionally worried
- Unable to concentrate

Any of these factors will affect your reactions, especially if you have to deal with an emergency.

Rest periods

You must observe mandatory rest periods in your daily driving schedule. On long journeys, try to plan them to coincide with a break at a motorway service area or refreshment stop. This is especially important at night, when a long journey can cause tiredness to set in.

It's illegal to stop anywhere on the motorway hard shoulder or slip roads for a rest. If you feel tiredness coming on open the windows, turn the heating down and get off the motorway at the next junction. When you get to a service area have a hot drink, wash your face (to refresh you) and walk round in the fresh air before driving on.

Bear in mind that a substantial meal accompanied by the warmth in the cab, the continual resonance of the engine and long uninterrupted stretches of road, especially at night, can produce the very conditions you need to avoid.

Regulations

Motorways are subject to specific rules and regulations that must be observed by all LGV drivers. Study those sections relating to motorways in *The Highway Code*. You also need to know, understand and obey motorway warning signs and signals.

Vehicle checks

Before driving on the motorway you should ensure that you carry out routine checks on your vehicle, especially considering the long distances and prolonged higher speeds involved.

Tyres

All tyres on your vehicle (and any trailer) must be in good condition.

Tyres can become very hot and disintegrate under sustained high-speed running. Check for excessive heat when you stop for a break.

Inspect inside and outside visible faces for signs of

- Wear
- Damage
- Bulges
- Separation
- Exposed cords

Make sure that your vehicle has the correct-sized wheels fitted. Smaller diameters will run faster and may overheat on longer journeys. Ensure that all tyres are suitable for the loads being carried.

If a tyre bursts or shreds you may be able to see this in your mirrors. If you see smoke from the tyres you should stop as soon as it's safe to do so.

Also, make a habit of checking the tyre pressures regularly.

Mirrors

Ensure that all mirrors are properly adjusted to give the best possible view to the rear. They should also be clean. The simple device of tying a piece of cloth to the mirror bracket cleans effectively as the air flow causes it to continually wipe the surface. Make sure that you tie it on tightly so that it doesn't work free and fly off.

Windscreen

All glass must be

- Clean
- Clear
- Free from defects

Keep all windscreen washer reservoirs topped up and the jets clear. Make sure that all wiper blades are in good condition.

Spray-suppression equipment

It's essential that you check all spray-suppression equipment fitted to the vehicle and any trailer before setting out, especially if bad weather is expected.

Instruments

Check all gauges and warning lights such as

- Anti-lock brakes (ABS)
- Air pressure
- Oil pressure
- Coolant
- Temperature
- Lights

Lights and indicators

To comply with the law all lights must be in working order even in daylight. Make sure that all bulbs, headlight units, lenses and reflectors are fitted, clean and function as intended.

High-intensity rear fog lights and marker lights (if fitted) must operate correctly. Indicator lights must operate and 'flash' within the specified frequency range. Reversing lights must either automatically operate by the selection of reverse gear or be switched on from the cab with a warning light to show when they're lit.

Fuel

Make sure that you either have enough fuel on board to complete the journey or have the facility (cash, agency card, etc.) to refuel at a service area.

Oil

The engine operates at sustained high speeds on a motorway so it's vital to check all oil levels before setting out. Running low can result in costly damage to the engine and could cause a breakdown at a dangerous location.

Coolant

The engine will be running for sustained periods so it's essential to check the levels of coolant in the system.

Joining a motorway

There are three alternative ways in which traffic can join a motorway. All these access routes will be clearly signed.

At a roundabout

The exit from a roundabout will be signposted. To prevent non-motorway traffic accidentally entering the system signs are displayed prominently.

Main trunk road becomes a motorway

There will be prominent advance warning signs so that prohibited traffic can leave the main route before the motorway regulations apply.

Via a slip road

Slip roads leading directly onto the motorway will be clearly signed to prevent prohibited traffic entering the motorway.

Effective observation Before joining the motorway from a slip road try to assess what traffic conditions are like on the motorway itself. You may be able to do this as you approach from a distance or if you need to reach the entry point by means of an over-bridge.

Get as much advance information as you can to help plan your speed on the slip road. You'll need to build up your speed and emerge safely onto the main carriageway.

Plan your approach and try to avoid having to stop at the end of the slip road. But if the motorway is extremely busy you may **have to** stop and filter into the traffic. Don't use the size of your vehicle to force your way onto the motorway. Use your mirror and signal as you pull out onto the main carriageway, if it's safe to do so.

A quick sideways glance may be necessary to ensure that you correctly assess the speed of any traffic approaching in the nearside lane. Don't

- Pull out into the path of traffic in the nearside lane if this will cause it to slow down or swerve
- Drive along the hard shoulder to 'filter' into the left-hand lane

There are a small number of locations where traffic merges onto the motorway from the right. Take extra care in these situations.

Making progress

Approaching access points

After passing a motorway exit there will often be an entrance or access point onto the motorway. Look well ahead and if there are vehicles joining the motorway

- Don't try to race them while they're on the slip road
- Be prepared to adjust your speed
- Move to the next lane, if it's safe to do so, to allow joining traffic to merge

Lane discipline

Keep to the left-hand lane unless overtaking slower vehicles. LGVs aren't allowed in the extreme right-hand lane on a three- or multi-lane motorway unless there are road-works or signs that indicate otherwise. On two-lane motorways LGVs are permitted to use the right-hand lane for overtaking.

Use the MSM/PSL routine well before signalling to move out. Don't start to pull out and then signal.

On a three- or four-lane motorway make sure that you check for any vehicle in the right-hand lane(s) that might be about to move back to the left. Most of the traffic coming up behind will be travelling at a much higher speed.

Look well ahead to plan any overtaking manoeuvre, especially given the effect a speed limiter will have on the power available to you.

Observe signs showing a crawler lane for LGVs. This will suggest a long, gradual climb ahead.

If a slow-moving oversized load is being escorted look for any signal the escort might give. They may permit you to move into the right-hand lane to pass the obstruction.

If a motorway lane merges from the right (in a few cases only) you should move over to the left as soon as it's safe to do so. At these specific locations no offence is committed if an LGV is initially travelling in the extreme right-hand lane. Move over to the left as soon as it's safe to do so.

Separation distance

On motorways you should allow

- Greater margins than on normal roads
- A safe separation distance

In good conditions you'll need at least

- One yard for every mph
- A two-second time gap

In poor conditions you'll need at least

- Double the distance
- A four-second time gap

In snow or icy conditions the stopping distances can be ten times those needed in normal dry conditions.

Seeing and being seen

Make sure that you start out with a clean windscreen, mirrors and windows. Use the washers, wipers and demisters to keep the screen clear. In poor conditions use dipped headlights.

Keep reassessing traffic conditions around you. Watch out for brake lights or hazard flashers that show the traffic ahead is either stationary or slowing down.

High-intensity rear fog lights should only be used when visibility falls below 100 metres (about 330 feet). They should be switched off when visibility improves, unless fog is patchy and danger still exists.

Motorway signs and signals

Motorway signs are larger than normal road signs. They can be read from greater distances and can help you to plan ahead.

Know your intended route. Be ready for the exit that you need to use and prepare for it in good time, well before you reach it.

Where there are major roadworks there may be diversions for LGVs in operation. Look for the yellow

- Square
- Diamond
- Circular

symbols combined with capital route letters. Follow the symbol on the route signs.

Signals

Warning lights show when there are dangers ahead such as

- Accidents
- Fog
- Icy roads

Look out for variable message warning signs advising

- Lane closures
- Speed limits
- Hazards
- Standing traffic ahead

You need to comply with advisory speed limit signs shown on the motorway hazard warning lights matrix.

Red light signals

If the red X signals show above your lane don't go beyond the red light.

- Be ready to comply with any signs that tell you to change lanes
- Be ready to leave the motorway
- Observe brake lights or flashing hazard warning lights, which show that there's stationary or very slow-moving traffic ahead

React in good time.

Lane control signals ahead

lane open

lane closed

move to left

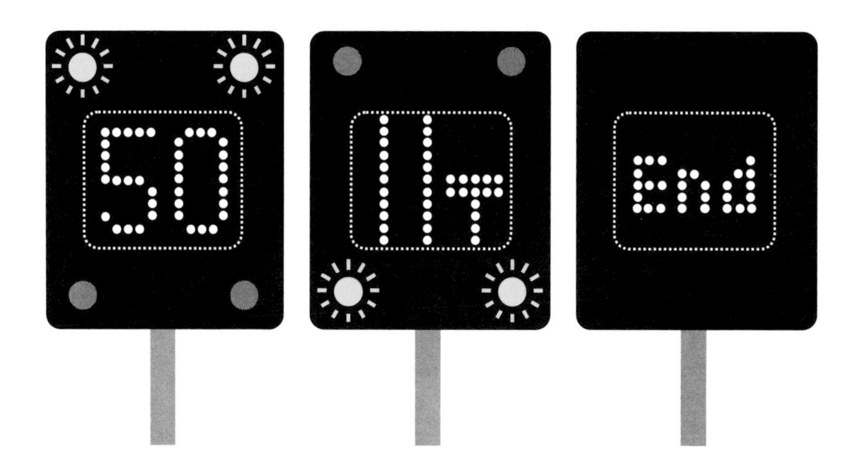

Weather conditions

Because of the higher speeds on motorways it's important to take into account any effects the weather may have on driving conditions. Listen to weather forecasts on the radio.

Rain

Visibility can be reduced by the spray thrown up by numbers of LGVs travelling at speed.

- Use headlights so that other drivers can see you
- Reduce speed when the road surface is wet. You need to be able to pull up in the distance that you can see to be clear
- Leave a greater separation gap. Use the four-second rule as a minimum
- Make sure that all spray-suppression equipment fitted to your vehicle is effective
- Take extra care when the road surface is wet after rain. The roads may still be slippery even if the sun is out

Crosswinds

Be aware of the effects strong crosswinds can have on other road users. Watch out especially

- After passing motorway bridges
- On elevated exposed sections
- When passing vehicles towing caravans, horse boxes, etc.

If you're driving a high-sided vehicle such as a

- Furniture removal van (pantechnicon)
- Box van carrying comparatively light merchandise
- Curtain-sided vehicle

take special notice of warnings for drivers of such vehicles. Avoid known problem areas such as viaducts, high suspension bridges, etc.

Motorcyclists are especially vulnerable to severe crosswinds on motorways so allow room when overtaking them. Check the nearside mirror to observe them after you've overtaken.

Ice or frost

In cold weather, especially at night when temperatures can drop suddenly, be on the alert for any feeling of lightness in the steering. This may suggest frost or ice on the road. Watch for signs of frost also along the hard shoulder. A warm cab can isolate you from the real conditions outside.

Motorways that appear wet may in fact be frozen. There are devices that fix onto the exterior mirror to show when the outside temperature drops below freezing.

Allow up to **ten** times the normal distance for braking in icy conditions. Remember, all braking should be carried out gently to reduce the risk of losing control.

Fog

If there's fog on the motorway you must reduce speed so that you can pull up in the distance that you can see to be clear. You should

- Slow down
- Use dipped headlights
- Use rear high-intensity fog lights if visibility is less than 100 metres (about 330 feet)
- Stay back
- Check your speedometer

Don't

- Speed up again if the fog is patchy. You could run into dense fog again within a short distance
- Hang onto the rear lights of the vehicle in front

Fog affects your judgement of speed and distance. You may be travelling faster than you realise.

Slow down.

Multiple pile-ups on motorways don't just happen – they're caused by drivers who

- Travel too fast
- Drive too close
- Assume there's nothing stopped ahead
- Ignore signals

You can't see well ahead in fog.

Be prepared to leave the motorway and watch out for signals. Be on the alert for accidents ahead. Watch out for emergency vehicles coming up behind, possibly on the hard shoulder.

Serious multiple accidents will continue to occur in future until all drivers get the message to

Slow down in fog.

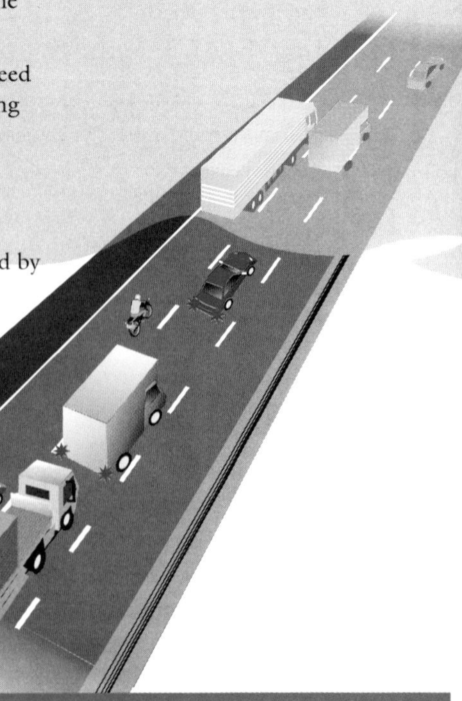

Contraflows and roadworks

Essential roadworks involving opposite streams of traffic sharing one carriageway are known as contraflow systems. The object is to permit traffic to continue moving while repairs or resurfacing take place on the other carriageway or lanes.

Red and white marker posts are used to separate opposite streams of traffic. The normal white lane-marking reflective studs are replaced by temporary yellow/green fluorescent studs.

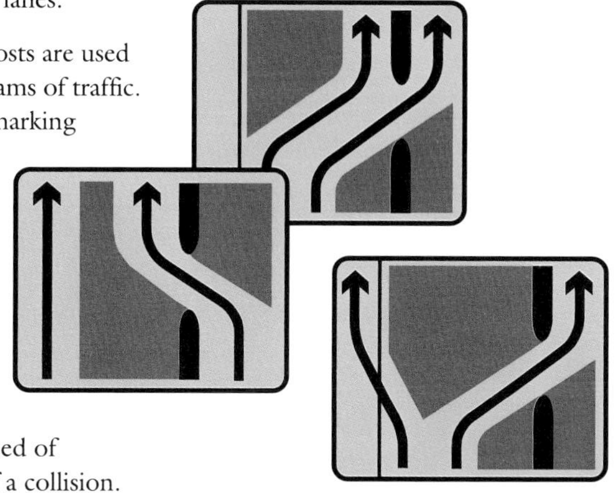

A 50 mph mandatory speed limit is usually imposed over the stretch affected. This means a closing speed of 100 mph in the event of a collision.

- Concentrate on what's going on ahead
- Don't let the activity on the other section distract you
- Don't exceed the speed limit
- Keep a safe separation distance from the vehicle in front
- Look well ahead to avoid the need to brake sharply
- Comply with advance warning signs, which indicate lanes that must not be used by LGVs (vehicles over 7.5 tonnes are prohibited)
- Avoid sudden steering movements or any sharp braking
- Don't change lanes if signs tell you to stay in your lane
- Don't speed up until you reach the end of the roadworks and normal motorway speed limits apply again

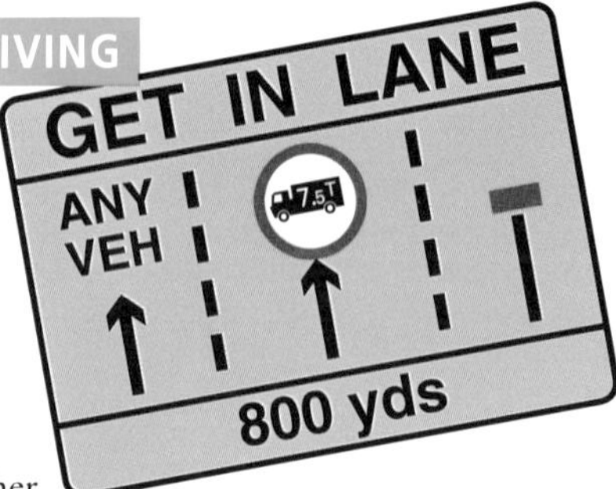

Accidents

Serious accidents can often occur when vehicles cross into the path of the other traffic stream in a contraflow.

- Keep your speed down
- Keep your distance
- Stay alert

Signs

Take notice of advance warning signs. Get into lane in good time. Avoid forcing your way in at the last moment.

Drivers of LGVs carrying oversized loads MUST comply with the advance warning notices. These will tell you either to leave the motorway or to stop and telephone the police and wait for an escort through the roadworks.

Breakdowns

If your vehicle breaks down in the roadworks section remain with the vehicle. These sections of motorway are usually under TV monitoring. A recovery vehicle, free within the roadworks section, will generally be with you soon.

Be on the lookout for broken-down vehicles causing an obstruction ahead.

Breakdowns on the motorway

- Try to get the vehicle as far to the left as possible
- Don't attempt even minor repairs on the motorway
- Operate the hazard warning flashers on the vehicle
- Make sure that the vehicle lights are on at night time, unless an electrical fault prevents this

If an object or part of any load falls onto the motorway from either yours or any other vehicle

- Don't attempt to recover it yourself
- Don't stand on the carriageway to warn oncoming traffic
- Without delay, use the nearest emergency telephone to call the police

The location of the nearest telephone will be indicated by the direction of the arrow on the marker posts at intervals along the edge of the hard shoulder. Don't cross the carriageway or any slip road to reach a telephone. Face the oncoming traffic while using the phone.

The motorway emergency telephones are free. You'll be connected directly to the motorway police control centre, who will then contact a recovery contractor.

Leaving the motorway

Progressive signs will show upcoming exits. At one mile you'll see

- The junction number
- The road number
- The one-mile indicator

Half a mile from the exit you'll see signs for

- The main town or city served by the exit
- The junction number
- The road number
- The half-mile exit

Finally, 300 yards (270 metres) before the exit there will be three countdown markers every 100 yards.

A driver of a vehicle travelling at 60 mph has only 60 seconds from the one-mile sign to the exit. So even at a speed of 50 mph there's still only 80 seconds from the one-mile sign to the actual exit.

Plan well in advance in order to be in the left-hand lane in good time. Large vehicles in the left-hand lane may prevent a driver in the second lane from seeing the one-mile sign, leaving very little time to move to the left safely.

You must use the MSM/PSL routine in good time before changing lanes or signalling. Assess the speed of traffic ahead in good time in order to avoid overtaking and then having to pull back in and reduce speed to leave at the next exit. Don't

- Pull across at the last moment
- Drive over the white chevrons that divide the slip road from the main carriageway

Occasionally there are several exits close together or a service area close to an exit. Look well ahead and plan your exit in good time. Watch out for other drivers' mistakes, especially those leaving it too late to exit safely.

Traffic queuing

At some locations traffic can be held up on the slip road. Look well ahead and be prepared for this. Don't queue on the hard shoulder.

Illuminated signs have been introduced at a number of such locations to give advance warning messages of traffic queuing on the slip road or in the first lane. Watch out for indicators and hazard warning flashers when traffic is held up ahead.

Use the MSM/PSL routine in good time and move to the second lane if you aren't leaving by such an exit.

End of the motorway

There are 'End of Motorway' regulation signs

- At the end of slip roads
- Where the road becomes a normal main road

These remind you that different rules apply to the road that you're joining. Watch out for signs advising you of

- Speed limits
- Dual carriageways
- Two-way traffic
- Clearways
- Motorway link roads
- Part-time traffic signals

Reduce speed

After driving on the motorway for some time it's easy to become accustomed to the speed. When you first leave the motorway, 40 or 45 mph can seem more like 20 mph. You should

- Adjust your driving to the new conditions as soon as possible
- Check the speedometer to see the real speed

Start reducing speed when you're clear of the main carriageway. Motorway slip or link roads often have sharp curves that need to be taken at lower speeds.

Look well ahead for traffic queuing at a roundabout or traffic signals. Be prepared for the change in traffic at the end of the motorway. Look out for

- Pedestrians
- Cyclists, etc.

Goods need to be delivered 24 hours a day all year round. With that in mind, you should employ safe driving techniques to ensure that you, your vehicle and the goods in your care always arrive safely at their destination with the minimum of delay. You'll need all your skills to achieve this objective during periods of bad weather.

It's essential that you take notice of warnings of severe weather such as

- Snow or blizzard conditions
- Floods
- Fog
- High winds

If an LGV becomes stranded the road may well be blocked for essential rescue and medical services. In the case of fog it could result in other vehicles behind colliding with the stranded vehicle.

Training and preparation are vital. Don't venture out in severe weather conditions without being properly prepared.

Your vehicle

Your vehicle must be in a fit and proper condition at all times. This means regular safety checks and strict observance of maintenance schedules.

Tyres

Check the tread depth and pattern. Examine tyres for cuts, damage and signs of cord visible at the side walls.

Brakes

It's essential that the brakes are operating correctly at all times. Any imbalance could cause a skid if the brakes are applied on a slippery surface.

Oil and fuel

Use the correct grades of fuel and oil for any extreme conditions.

Prolonged hot weather will place additional demands on the lubricating oil in engines and turbochargers. In extremes of cold it will be necessary to use diesel fuel with 'anti-waxing' additives to prevent fuel lines freezing up.

In excessively dusty conditions, which can be encountered on construction sites, quarries, etc., you should strictly follow schedules relating to filter changes.

Icy weather

Ensure the whole of the windscreen is cleared before attempting to move off in frosty conditions.

If you're driving at night time be alert for any drop in temperature that could cause untreated roads to become icy. If the steering feels light you're probably driving on a frozen road surface, so ease your speed as soon as it's safe to do so. All braking must be gentle and over much longer distances, especially when driving articulated vehicles or those with a trailer.

You'll have to allow more time for the journey because overall speeds will need to be lower. Also, keep a safe separation distance from any vehicle ahead. Allow **ten** times the normal stopping distance.

Drive sensibly and allow for the fact that other road users might get into difficulties. Avoid any sudden braking, steering or acceleration.

Heavy rain

Ensure that the wipers are clearing the windscreen properly – you'll need to be able to see clearly ahead. Make sure that the screen is also demisted efficiently and that the washer containers are filled with suitable fluid, especially in winter conditions.

Allow at least twice as much separation distance as you would in dry conditions. If you must brake, do it while the vehicle is stable and preferably travelling in a straight line. Avoid sudden or harsh braking.

Obey advisory speed limit signs on motorways.

Other road users will have more difficulty seeing when there's heavy rain and spray. Make sure that all spray-suppression equipment on your vehicle is secure and operating.

Don't use high-intensity rear fog lights unless visibility is less than 100 metres (330 feet).

Construction sites

Heavy rain can turn constructions sites into quagmires. Take care when driving on off-road gradients, or when getting down from the cab.

If the vehicle is fitted with a switch for locking up the differential mechanism on the drive axle (the 'diff-lock'), engage it. This will ensure that the power is transmitted to all driven wheels and will assist traction by eliminating wheelspin. Remember to disengage the diff-lock as soon as you return to normal road surfaces again.

It's an offence to deposit mud on the roadway to the extent that it could endanger other road users. This may involve hosing down the wheels and undergear of your vehicle before it leaves such a site. You should also check between double wheels before leaving the site for any large stones or building bricks wedged between the tyres. Such objects can fly out at speed, with serious consequences for following traffic.

Snow

Falling snow can reduce visibility dramatically. Use dipped headlights and reduce your speed. Allow a much greater stopping and separation distance – up to **ten** times the stopping distance on dry roads.

Road markings and traffic signs can become obscured by snow. Take extra care at junctions.

Deep snow as a result of drifting in high winds can often lead to the closure of high-level roads. Don't attempt to use such roads if

- Broadcasts tell LGV drivers to avoid those routes
- Warning signs indicate that the road is closed to LGVs or other traffic
- Severe weather conditions are forecast

Some rural roads in exposed places have marker posts at the side of the road that will give a guide to the depth of the snow.

In prolonged periods of snow the fixing of snow chains to driven wheels will often prove to be of value. Remember, a stranded LGV could

- Prevent snow ploughs clearing the route
- Delay emergency vehicles
- Block the road for other road users

Ploughs and gritting vehicles

Don't attempt to overtake a snow plough or gritting vehicle. You may find yourself running into deep snow or skidding on an untreated section of roadway which these maintenance vehicles could have cleared or treated had you followed on behind them. Keep well back from gritting vehicles. If they're on the road there could be bad weather on the way.

Deep snow

If your vehicle becomes stuck in deep snow engage the diff-lock (if one is fitted) to regain forward traction. Switch it off as soon as the vehicle is moving and before attempting a turn.

Another technique for freeing a vehicle stuck in the snow is to use the highest gear, which will provide traction. Then try alternating between reverse and the forward gear until forward motion is possible. Avoid continual revving in a low gear. This will only result in the drive wheels digging an even deeper rut.

It's often helpful to keep a couple of strong sacks in the cab to put under the drive wheels if the vehicle becomes stuck. A shovel is also handy if the journey is likely to involve crossing areas where snow is known to be a hazard during the winter.

When operating independent retarders care must be taken on the descent of snow-covered gradients. The retarders could cause the rear wheels to lock. Some retarders are managed by the ABS to help avoid this problem.

Fog

- Don't drive in dense fog if you can postpone the journey
- Avoid driving at all in night-time fog
- Avoid starting a journey that might need to be abandoned because it's too dangerous to proceed any further

The options for finding a safe place to park an LGV off the road in dense fog are limited. You must not leave an LGV on or near a public road where it would become a danger to other road users. Certainly don't leave any LGV or trailer without lights where it could be a danger to other road users.

Lights

Use dipped headlights in any reduced visibility. You need to see and be seen.

Use rear high-intensity fog lights and front fog lights when visibility is less than 100 metres (330 feet). These must only be capable of operating when the dipped headlights are switched on. Switch off front and rear fog lights when visibility improves above 100 metres (330 feet), but beware of patchy fog.

In all cases but especially in poor weather conditions, keep all lenses and reflectors clean and ensure that all lights are working correctly.

Reflective studs and markings

Reflective studs are provided on dual carriageways to help drivers in bad visibility. The colours of reflective studs are

- **Red** on the left-hand edge of carriageways
- **White** to indicate lane markings
- **Green** at slip roads and lay-bys
- **Amber** on the right-hand edge of carriageways and the centre reservation

On some rural roads there are black and white marker posts with red reflectors on the left-hand side and white reflectors on the right-hand side of the road.

The continuous white line between the left-hand lane and the hard shoulder and at the left-hand edge of some trunk roads incorporates a 'rumble' strip. This produces a vibration designed to warn drivers when their vehicle crosses the line.

In fog don't

- Drive too close to the centre of the road
- Confuse centre lines and lane markings
- Drive without using headlights
- Speed up because the fog appears to thin
- Use full beam when following another vehicle – the shadows will make it difficult for the driver ahead to see

A large vehicle travelling ahead of you may temporarily displace some of the fog, making it appear clearer than it really is. But then again, in a larger vehicle you may be able to see ahead over some low-lying fog. Don't speed up in case there are smaller vehicles in front that may be hidden from view.

Slow down.

- Don't speed up if the fog appears to thin. It could be patchy and you could run into it again
- Keep checking the speedometer to see your true speed. Fog can make it difficult to judge speed and distance

Stay back.

- Keep a safe separation distance from any vehicle ahead
- Don't speed up if a vehicle appears to be close behind
- Only overtake if you can be **sure** the road ahead is clear

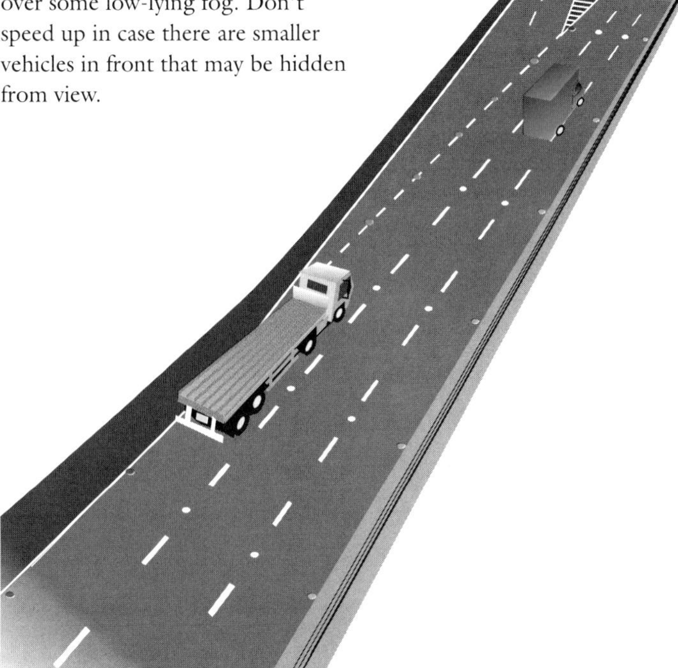

High winds

In severe weather conditions you should plan your journey well in advance (often 24 hours). Listen to, watch or read the weather forecast, especially if you're the driver of

- A high-sided vehicle (removal vans, long wheel-base box vans, etc.)
- A vehicle with a curtain-side body or trailer
- A vehicle transporting portable buildings, etc. with large flat surfaces susceptible to wind pressures
- A vehicle towing a horse box
- An unladen van of any description

Take notice if your route includes any locations that are frequently subjected to high winds such as

- High-level bridges or roads
- Exposed viaducts
- Exposed stretches of motorway

Watch out for signs indicating high winds. Also, beware of fallen trees or damaged branches that could fall on your vehicle.

Take notice of advance warnings and always remember that

- The route may be closed to certain LGVs
- There may be additional delays due to lane closures. This is done on high-level bridges to create empty 'buffer' lanes, which cope with vehicles that are blown off course into the next lane
- You may need to use an alternative route
- If you ignore the warnings, your vehicle and its load could be affected by the strong winds and could place you and other road users in danger

Bear in mind that ferry sailings are likely to be affected by gale force winds, resulting in delays or cancellation.

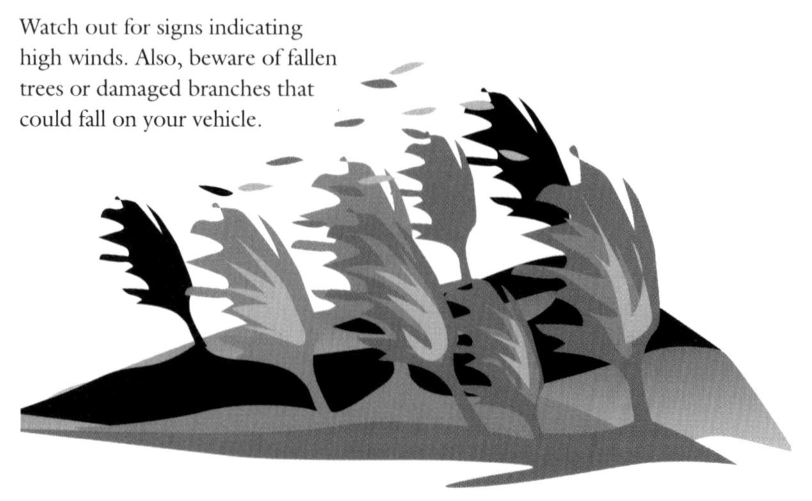

Other road users

In windy conditions other road users are likely to be affected when

- Overtaking your vehicle
- You overtake them

Check the nearside mirror(s) as you overtake to ensure that they still have control of their vehicle. In addition, be on the alert for vehicles or motorcyclists 'wandering' into your lane.

Don't ignore warnings of severe winds. If your vehicle is blown over you could delay the emergency services from reaching an even more serious incident.

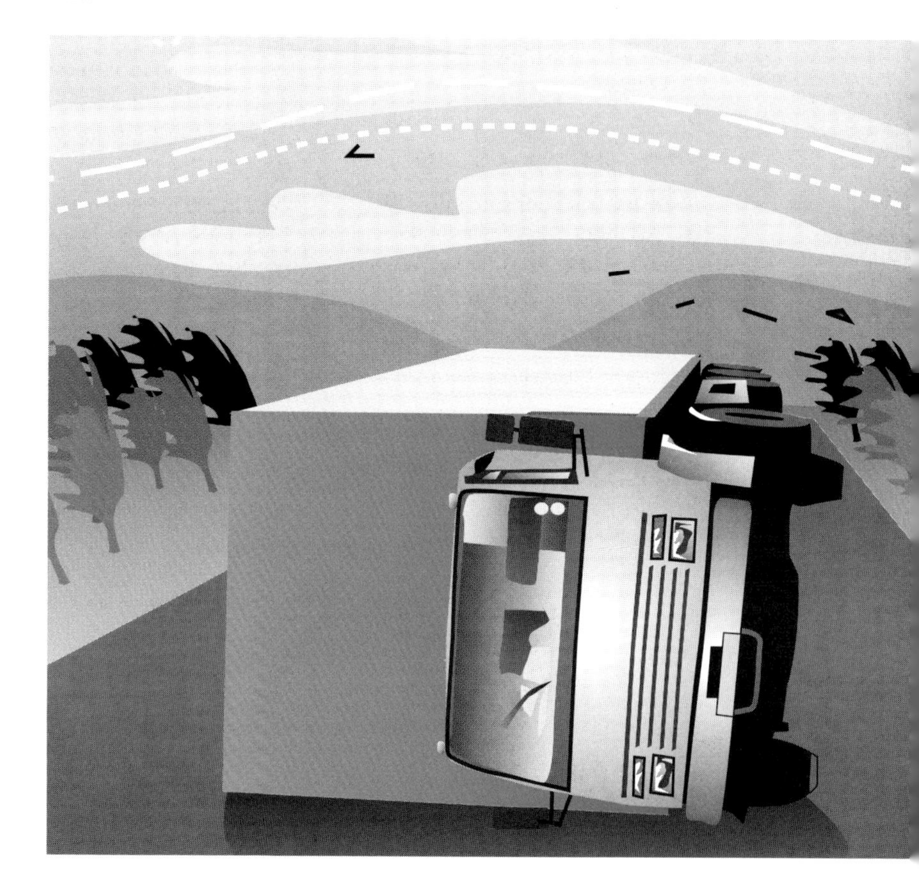

You should drive at all times with anticipation and awareness. By acting in this way you lessen the risk of being involved in an accident.

It's important to recognise the effects your vehicle can have on more vulnerable road users such as cyclists, pedestrians and motorcyclists. An LGV can create a vacuum effect when travelling at speed. Pedestrians near the edge of the kerb and cyclists are especially vulnerable to the danger of being drawn under the wheels of your vehicle or any trailer. You should anticipate at all times the actions of other road users around you.

- Concentrate
- Stay alert
- Be fully fit
- Observe the changes in traffic conditions
- Plan well ahead
- Drive at a safe speed to suit the road and traffic conditions
- Keep your vehicle in good mechanical condition
- Ensure that the load is securely stowed
- Drive safely and sensibly
- Avoid the need to rush
- Don't act hastily

If your vehicle is involved in a road traffic accident

You must stop.

It's an offence not to do so.

At an accident scene

If you're the first or among one of the first to arrive at the scene of an accident your actions could be vital. It's essential to

- Warn other traffic approaching the scene by means of hazard warning flashers, beacons, cones, advance warning triangles, etc.
- Reduce the risk of fire by making sure that all naked lights – cigarettes, etc. – are extinguished
- Make sure that someone phones 999, giving details of any injury or danger to other road users
- Protect injured persons from any danger from traffic, hazardous materials, etc. It may well be best to keep them still until the emergency services arrive
- Be especially careful about moving any casualties – incorrect handling could cause more injury or even prove fatal
- Move any apparently uninjured persons away from the vehicle(s) to a place of safety
- Give First Aid if anyone is unconscious
- Check for the effects of shock. A person may appear to have no injuries but may be suffering from shock
- Keep casualties warm but give them nothing to drink
- Give the **facts** (not assumptions, etc.) to the ambulance crew when they arrive

Accidents on the motorway

Because of the higher speeds on motorways and increased danger of an accident becoming a serious incident, it's essential to inform the motorway police and emergency services as quickly as possible.

- Use the nearest emergency telephone if no mobile telephone is available
- Don't cross the carriageway to get to an emergency telephone
- Try to warn oncoming traffic, if possible, without placing yourself in danger
- Move any uninjured people well away from the main carriageway and onto an embankment, etc.
- Be on the alert for emergency vehicles approaching the incident along the hard shoulder

Hazardous materials

If a road traffic accident involves a vehicle displaying either a hazard warning information plate or a plain orange rectangle

- Give the emergency services as much information as possible about the labels and any other markings
- Contact the emergency telephone number on the plate of a vehicle involved in any spillage, if a number is given
- Keep well away from such a vehicle unless you have to save a life
- Beware of any dangerous liquids, dusts or vapours – no matter how small the concentration may appear to be. People have received extremely serious injuries as a result of a fine spray of corrosive fluid leaking from a pinhole puncture in a tank vessel

Examples of various hazard labels are shown in Part Six.

Fire

Fire can occur on LGVs in a number of locations

- Engine
- Load
- Transmission
- Tyres
- Fuel system
- Electrical circuits

It's vital that any fire is tackled without delay. A vehicle and its load can be destroyed by fire within an alarmingly short period of time.

If fire is suspected or discovered, and if danger to others is to be avoided, it's essential to

- Stop as quickly and safely as possible
- Get all individuals out of the vehicle
- Either dial 999 or get someone else to do it immediately
- Tackle the source with a suitable fire extinguisher

If the fire involves a vehicle carrying hazardous materials

- The driver must have received training in dealing with such an emergency
- Specialist fire-fighting equipment must be available on the vehicle
- Keep all members of the public and other traffic well away from the incident
- Isolate the vehicle to reduce danger to the surrounding area
- Ensure that someone contacts, without delay, the emergency telephone number given on either the hazard warning plate or the load documents
- Warn approaching traffic

Stay calm.

React promptly.

Fire extinguishers

You should be able to recognise the various types of fire extinguisher and know which fires they're intended to tackle. For example, it's dangerous to tackle a fuel fire with a water type of soda-acid fire extinguisher since this may only spread the fire further.

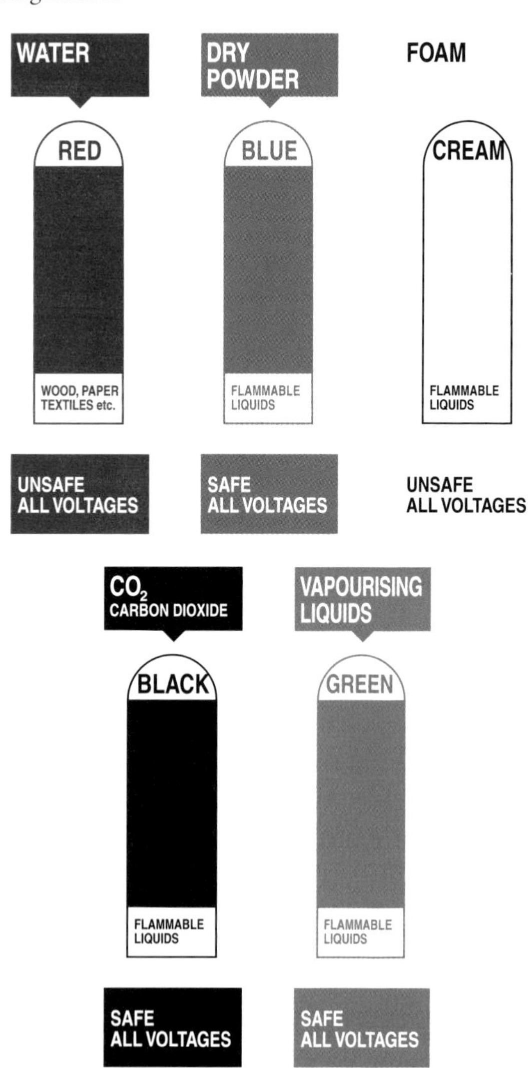

Most extinguishers are intended to smother the source of the fire by either the action of an inert gas or a dry powder. Try to isolate the source of the fire. If at all possible

- Disconnect electrical leads
- Cut off the fuel supply

Don't open an engine housing wide if you can direct the extinguisher through a small gap. Also, avoid operating a fire extinguisher in a confined space.

Vehicles carrying high-risk materials are subject to detailed emergency procedures which must be followed.

First Aid

Regulations require many LGVs carrying chemicals, etc. to carry First Aid equipment. Even if you don't have to carry a kit by law, it's sensible for every LGV driver to have a First Aid kit available.

You should really consider doing First Aid training. Someday it could save a life. There are courses available from the

- St John Ambulance Association and Brigade
- St Andrew's Ambulance Association
- British Red Cross Society

If you don't manage to attend a First Aid training course the following information may be of some assistance.

Unconscious victims

It's vital that action is taken within the first three minutes of an incident if a casualty is to be saved. The easiest way to remember the action to take is to think of the letters **ABC**

- **A**irway must be cleared of any obstruction and kept open
- **B**reathing must be established and maintained
- **C**irculation must be maintained and severe bleeding stopped

Breathing stopped Get breathing started again by

- Removing any obstruction such as false teeth, chewing gum, etc.
- Keeping the victim's head tilted backwards

Breathing and colour should improve. If not

- Place a clean piece of material, such as a handkerchief, over the injured person's mouth
- Pinch the casualty's nostrils together and blow into the mouth until the chest rises
- Let your mouth surround the mouth and nose of small children and babies and blow very gently
- Take your mouth away and wait for the chest to fall
- Withdraw, then repeat regularly once every four seconds until breathing restarts and the casualty can breathe without help

Don't give up!

Never assume someone is dead. Keep giving mouth-to-mouth resuscitation until medical help is available.

Unconscious and breathing If you suspect a head injury, avoid moving the casualty if at all possible until medical help is available. Only move the casualty if they're in danger of further injury. Don't attempt to remove a safety helmet unless it's absolutely essential, otherwise an injured motorcyclist, for example, could sustain even more serious injuries.

It's vital to obtain skilled medical help as soon as possible. Make sure that someone dials 999.

Bleeding To stem the flow of blood put firm pressure on the wound without pressing on anything that may be caught in or projecting from the wound.

As soon as practicable secure a pad to the wound with a bandage or length of cloth. Use the cleanest material available.

If a limb is bleeding, but not broken, raise it to lessen the flow of blood. Any restriction of blood circulation for more than a short period of time may result in long-term injury.

Dealing with shock

The effects of trauma may not be immediately obvious. Prompt treatment can help to minimise the effects of shock.

- Don't give victims anything to drink until medical advice is available
- Reassure the victim confidently and keep checking them
- Keep any casualties warm and make them as comfortable as possible
- Try to calm a hysterical person by talking to them in firm, quiet tones
- Make sure that shock victims don't run into further danger from traffic
- Avoid leaving anyone alone
- Avoid unnecessary movement of a casualty

Electric shock

A vehicle can come into contact with overhead cables or electrical supplies to traffic bollards, traffic lights or street lighting standards as a result of an accident. Make a quick check before attempting to pull someone from a vehicle in such cases.

Don't touch any person who is obviously in contact with a live electric cable unless you can use some non-conducting item, such as a dry sweeping broom, length of wood, etc.

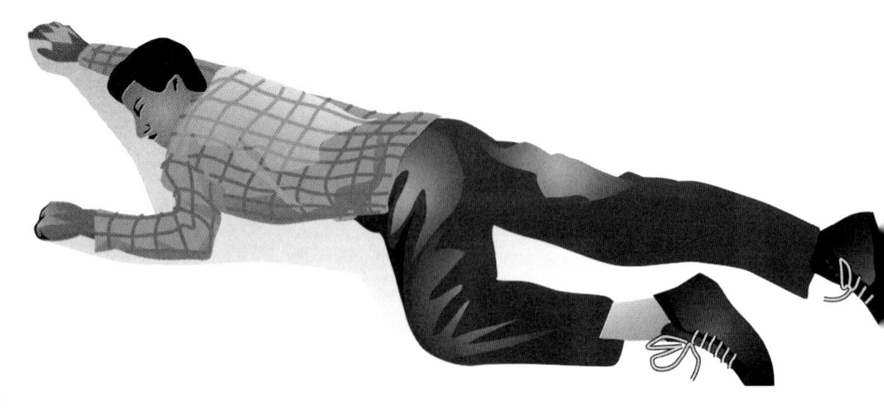

Reporting your accident

If your vehicle is involved in a road traffic accident STOP – it's an offence not to do so. You must

- Inform the police as soon as possible, or in any case within 24 hours, if
  - there's injury to any person not in your vehicle
  - damage is caused to another vehicle or property and the owner is either not present or can't be found easily
  - the accident involves any of the animals specified in law
- Produce your insurance documents and driving licence, and give your name and address to any police officer who may require it
- Give these details to any other road user involved in the accident if they have reasonable grounds to request them
- If you're unable to produce your documents at the time you must report the accident to the police as soon as possible, or in any case within 24 hours

The police may require you to produce your documents within seven days at a police station of your choice (five days in Northern Ireland) or as soon as is reasonably possible if you're on a journey that takes you out of the country at the time and you can't produce the documents within the seven days specified. You must

- Exchange particulars with any other driver or road user involved in the accident
- Obtain names and addresses of any witnesses who **saw** the accident

You should make a note of

- The time
- The location
- Street names
- Vehicle registration numbers
- Weather conditions
- Lighting (if applicable)
- Any road signs or road markings
- Road conditions
- Damage to vehicles or property
- Traffic lights (colour at the time)
- Any indicator signals or warning (horn)
- Any statements made by the other party or parties
- Any skid marks, debris, etc.

Tyre failures

Many LGV breakdowns involve tyre failures or 'blow-outs'. Not only are these dangerous in themselves, by causing loss of control, but the resulting debris also presents a common hazard to other road users.

Front wheel blow-outs

A sudden deflation of the front tyre on an LGV can result in a loss of steering control.

- Keep firm hold of the steering wheel
- You should always be aware of anything on your nearside
- Signal to move to the left
- Try to steer a steady course to the nearside (or hard shoulder on the motorway)
- Reduce speed gradually to avoid any harsh braking
- Try to bring the vehicle to rest under control and as far to the left as possible
- If you have a warning triangle place it behind the vehicle and operate the hazard flashers if the vehicle is causing an obstruction

Avoid sharp braking and excessive steering movements. You should be able to bring the vehicle to rest safely by reducing the risk of skidding.

Rear wheel blow-outs

If a rear tyre on either the vehicle or a trailer deflates the effects may not be quite so severe. On a large vehicle this may not be immediately obvious to you if it's a multi-axle trailer. Keep the trailer under observation at all times during a journey.

Lost wheels

Regular maintenance is essential to help prevent wheels becoming detached during use. When the wheels have been removed and replaced for any reason it's important to re-check the wheel nuts shortly after their initial tightening. Check the wheel fixings regularly during use, preferably as part of your inspection routine prior to starting any journey.

It's essential that wheel fixings are tightened to the torque specified by the vehicle manufacturer. You should also use a torque wrench that's frequently calibrated.

Further information is given in British Standard Code of Practice for the selection and care of tyres and wheels for commercial vehicles. This has been developed with the support and involvement of the major transport operators' associations. The relevant reference number is BS AU 50: Part 2: Section 7a: 1995, and it's available from the British Standards Institution, 389 Chiswick High Road, London W4 4AL.

PART FOUR PREPARING FOR THE DRIVING TEST

This part looks at how to prepare for the practical LGV driving test.

The topics covered

- About the driving test
- Theory into practice
- How to apply for your test
- Before attending your test
- Legal requirements
- At the test centre
- The official syllabus

When taking the LGV practical driving test you should aim for a professional standard. You'll pass if your examiner sees that you can

- Drive safely to a high standard
- Show expert handling of controls
- Carry out the set exercises accurately and under control
- Demonstrate through your driving that you have a thorough knowledge of *The Highway Code* and other matters concerning vehicle safety

Does the standard of the test vary?

No. All examiners are trained to carry out the test to the same standard. Test routes are

- As uniform as possible
- Include a wide range of typical road and traffic conditions

You should have the same results from different examiners or at different LGV driving test centres.

Are examiners supervised?

Yes, they're closely supervised. A senior officer may sit in on your test if there are two or more passenger seats in your vehicle.

Don't worry about this. The supervising officer won't be examining you, but will be checking that the examiner is carrying out the test properly. Just carry on as if she or he wasn't there.

Can anyone accompany me on the test?

Due to lack of seats this isn't always possible. If there are three or more seats in the cab of your vehicle, and provided a DSA supervisor isn't intending to observe the test, your instructor is allowed to be present but can't take any part in the test.

How should I drive during the test?

Drive in the way that your instructor has taught you.

If you make a mistake, try not to worry. It might be minor and may not affect the result of the test. Your examiner will be looking for a high overall standard. Don't worry about one or two minor mistakes.

What will my examiner want from me?

Your examiner will want you to drive safely to a high standard under various road and traffic conditions. You'll be

- Given directions clearly and in good time
- Asked to carry out set exercises

Due to the higher level of engine noise in the cab your examiner will make sure that you're able to hear the directions clearly.

Your examiner will be understanding and sympathetic. She or he will try to put you at your ease to help you to do your best.

What will the test consist of?

Apart from general driving, the test will include

- Reversing within a marked area into a restricted opening
- A braking exercise
- A gear-changing exercise, if you're driving a manual vehicle
- Moving off on the level, at an angle, uphill and downhill
- Demonstrating the uncoupling and recoupling procedure, if you're taking your test with a trailer

Three of the exercises are always carried out at the test centre. These are the

- Reversing exercise
- Braking exercise
- Uncoupling and recoupling exercise

The rest of the exercises will take place during the road section of the test.

During the reversing exercise your examiner will remain outside the vehicle.

Your examiner will join you in the cab before explaining the braking exercise to you. She or he will watch your handling of the controls as you carry out the exercise.

This exercise will be carried out before you leave the test centre. If your vehicle doesn't pull up satisfactorily your examiner may decide not to continue the test, in the interest of safety.

What if I don't understand?

Your examiner will be as helpful as possible and will explain what's required. Before the exercises you'll be shown a diagram. This will make it easier to understand what's required. You'll then be asked to carry out the exercise. If you aren't sure, ask. Your examiner won't mind explaining again.

How long will the test last?

About 90 minutes.

When will I be ready for my test?

When you've reached the standards set in this book – not before. You should ensure that you receive good instruction, together with as much practice as you can.

How will I know when I'm ready?

You're ready for your practical test when you're driving

- Consistently well
- With confidence
- In complete control
- Without assistance and guidance from your instructor

Most people fail because they haven't had enough instruction and practice. Make sure that all aspects of the syllabus for learning to drive an LGV are covered (see pages 146–56).

Special circumstances

You'll have had to pass a medical in order to obtain your provisional licence. Your doctor will have had to declare if there's any reason why you wouldn't have full control of a large vehicle. There may be circumstances when adaptations to a vehicle may overcome a particular disability.

To make sure that enough time is allowed for your test it would help the DSA to know

- If you're restricted in any way in your movements
- If you have any disability that might affect your driving

Please include this information when you apply for your test.

Your examiner may wish to talk to you about your disability and any adaptations you may have fitted to your vehicle.

Language difficulties

If you have difficulty speaking or understanding English you can bring an interpreter with you. Remember, the vehicle must have enough seats. The interpreter must be 16 years or over and must not be your instructor.

Before you take your practical LGV driving test you'll have to pass an LGV theory test. You must satisfy your examiner that you've **fully understood** everything that you learned for the theory test.

The various aspects include the knowledge of

- The height, weight, width and length of your vehicle. This enables you to drive on roads with a full knowledge of any restrictions that might apply to your vehicle
- Rules on drivers' hours and rest periods, so that you can obey the legal requirements and are fit to drive safely
- Braking systems and speed limiters. You should be fully aware of how your brakes work and the importance of using them effectively
- The restricted view you have around your vehicle due to its size and dimensions. This size will also affect other road users' view – you must be aware of this
- Faults on your vehicle and being able to recognise and report any defects
- The factors relating to loading a vehicle safely and securely
- The effect of wind on your vehicle and on other road users around you
- The dangers of splashing spray or mud on other road users when overtaking them
- The course you have to take when turning, in order to allow for the length or overhang of your vehicle
- The correct actions to take if you're involved in or arrive at an accident
- Reducing the risks when overtaking other road users
- The dangers of leaving the cab of your vehicle on the offside

If you've passed your LGV theory test you'll have shown that you've taken the time to learn the basic aspects of becoming an LGV driver.

At the start of the drive in your practical test your examiner will ask you to follow the road ahead, unless asked to turn or the traffic signs direct you otherwise. From this point you should be able to demonstrate your understanding of the topics covered in the theory test.

The application form

You can obtain an application form (DL26) from a DSA Area Office or an LGV driving test centre. Look at the guidance notes carefully, especially those that refer to vehicle categories. You must have a provisional licence for the category of vehicle that you're going to drive.

Make sure that you give all the particulars asked for on the application form. If you miss anything out it could delay the date of your test.

Don't forget to send your fee. You may do this by sending a cheque or postal order. Make sure that it's crossed and made payable to the Driving Standards Agency. If you send a postal order keep the counterfoil.

Please don't send cash.

Send your application form to your DSA Area Office. The address can be found at the back of this book.

Trainer booking

If you're learning to drive with a training organisation they'll normally book your test for you. An arrangement with the Area Office allows them to book and pay for test appointments in advance. This enables them to arrange courses to culminate with a test appointment.

If you're a trainer and are interested in this scheme contact your Area Office for more details.

Booking by credit or debit card

You must be the card-holder. If you aren't then the card-holder must be present when a booking is made on the telephone.

When you telephone make sure that you have the completed application form with you. The booking clerk will want to know

- Your driver number, shown on your licence
- The type of test you wish to book
- Your personal details (name, address, day/evening telephone numbers)
- Unacceptable days or periods
- Any special circumstances, such as being accompanied by an interpreter
- Your credit card number and expiry date (and the issue number when using Switch)

If you use this service the booking clerk will be able to offer you an appointment over the telephone. You should receive written notification confirming the appointment within a few days.

The credit card telephone number for your area is shown on the application form and in the list of DSA Area Offices at the back of this book.

Saturday and evening tests

Saturday and weekday evening tests are available at some driving test centres. The fees for these are higher than for a driving test during the normal working hours on weekdays.

Evening tests are available during the summer months only.

You can get details from

- DSA Area Offices
- Driving test centres
- Your instructor

Your test appointment

Your DSA Area Office will send you a notification of your appointment, which is the receipt of your fee. Take this with you when you attend your test. The notification will include

- The time and place of your test
- The address of the driving test centre
- Other important information

If you haven't received notification after two weeks contact the Area Office.

Postponing your test

Contact the DSA Area Office where you booked your test if

- The date or time of your appointment isn't suitable
- You want to postpone or cancel your test

There's a cancellation date on your appointment card. If you wish to cancel your test you should do so before this date or you'll lose your fee.

Documents

You must have applied for and received a provisional licence in the category that you wish to take your test. Make sure that you have your provisional driving licence and your theory test pass certificate with you. Check that you've signed your licence – your test might be cancelled if you haven't done this. If you have a photo licence you must bring the 'counterpart' which is part of the licence.

If your licence does not show your photograph you must also bring a document which does, such as a passport or a workplace identity card. You should check with your instructor.

Photographic identification

The following are acceptable

- **a photo card licence issued by the DVLA**. If you have this type of licence you do not need to bring additional photographic identity
- **your passport**. This doesn't have to be a British passport
- **cheque guarantee card or credit card** with a signature and photograph of the candidate
- **an employer's identity or workplace pass** with the candidate's photograph and their name or signature or both
- **a Trades Union Card** with the candidate's photograph and signature
- **a Student Union Card** with reference to either the NUS or an educational establishment/course reference number and displaying the candidate's photograph or signature or both
- **a School Bus Pass** with the name of the issuing authority and a photograph and signature of the candidate
- **a card issued in connection with the sale and purchase of reduced price railway tickets** issued by a Railway Authority or other authorised body to purchase a reduced price railway ticket (e.g. a Young Person's Railcard)
- **a Gun Licence**, including a Firearm or Shotgun Certificate which bears the photograph and signature of the candidate
- **a Proof of Age Card** issued by the Portman Group with the candidate's photograph and signature
- **a Standard Acknowledgement Letter (SAL)** issued by the Home Office with photograph and signature of the candidate

If you don't have any of these you can bring a photograph, together with a statement that it's a true likeness of you. This can be signed by any of the following

- an LGV instructor who is included on the DSA register of approved LGV instructors
- an Approved Driving Instructor (Cars)
- DSA certified motorcycle instructor
- Member of Parliament
- medical practitioner
- local authority councillor
- teacher (qualified)
- Justice of the Peace
- civil servant (established)
- police officer
- bank official
- minister of religion
- barrister or solicitor
- Commissioned Officer in Her Majesty's Forces

I ....
(name of certifier), certify that this is a true likeness of ....
who has been known to me for
.... (number) months/years in my capacity as ....
Signed ....
Date ....
Daytime phone number ....
ADI/CBT/LGV instructor no. .....

If you don't bring these documents with you on the day you won't be able to take your test and you will lose your fee. If you have any queries about what photographic evidence we will accept, contact the national enquiry line.

No photo

No licence

No test

Your test vehicle

Make sure that the vehicle you intend to drive

- Is legally roadworthy and has a current certificate issued by the Vehicle Inspectorate
- Is fully covered by insurance for its present use and for you to drive
- Is unladen and doesn't have any large advertising boards fixed to the flat bed of the vehicle
- Is in the category in which you want to hold a licence
- Has ordinary L plates visible to the front and rear (or D plates, if you wish, when driving in Wales)
- Isn't being used on a trade licence
- Has a seat in the cab for the examiner
- Has enough fuel, not only for the test (at least 20 miles) but also for you to return to base

You should check the vehicle's

- Stop lamps
- Direction indicators
- Lenses and reflectors
- Mirrors
- Brakes
- Tyres
- Exhaust system (for excessive noise)
- Windscreen and washers
- Wipers

Make sure that the cab is clean and free from any loose equipment.

Change of vehicle

Please let the Area Office know if you want to bring a different vehicle from the one described on your application form. This will avoid unnecessary delay when you arrive for your test.

Minimum test vehicles (MTVs)

The minimum test vehicle (MTV) for

- **Category C1** is 4 tonnes maximum authorised mass (MAM) and capable of a speed of 80 kph (50 mph)
- **Category C1 + E** is 4 tonnes MAM and capable of a speed of 80 kph (50 mph) with a trailer of at least 2 tonnes MAM and at least 8 metres overall length
- **Category C** is 10 tonnes MAM, at least 7 metres in length and capable of 80 kph (50 mph). This doesn't apply to the tractive unit of an articulated vehicle
- **Category C + E** is either
  - an articulated vehicle of 18 tonnes MAM, at least 12 metres in overall length and capable of 80 kph (50 mph); or
  - a drawbar outfit, which is a combination of a category C vehicle with a trailer at least 4 metres platform length. The trailer must have a MAM of at least 4 tonnes and be suitably braked and coupled. The combined weight must be at least 18 tonnes, and the combined length must be at least 12 metres

You must show that you're competent to drive the vehicle in which the test is being conducted without danger to, and with due consideration for, other persons on the road. The legal requirements state that you must be able to

- Start the engine
- Move off straight and at an angle
- Maintain a proper position in relation to other vehicles
- Overtake and take an appropriate course in relation to other vehicles
- Turn left and right
- Stop within a limited distance, under full control
- Stop normally and bring the vehicle to a rest in an appropriate part of the road
- Drive the vehicle forward and backward, and whilst driving the vehicle backward steer it along a permitted course so that it enters a restricted opening and then bring it to rest in a predetermined position
- Indicate your intended actions by appropriate signals at appropriate times
- Act correctly and promptly in response to all signals given by any traffic sign, any person lawfully directing traffic or any other person using the road
- Uncouple and recouple your trailer (if appropriate)

Your examiner won't carry out an eyesight test at the beginning of the practical test. You should already have met the requirements before your provisional licence was issued.

Before the drive

Make sure that you arrive in good time. This will allow you to make use of the toilet facilities and to get yourself settled before the test starts.

The test will take about 90 minutes so make sure that you won't exceed the number of hours that you're allowed to drive by law.

Your examiner will call your name and ask to see your licence and photo ID. Make sure that you've signed your licence. She or he will then ask you to sign a declaration that your insurance is in order. The test won't be conducted if you're unable to do so. When you've done this you'll be asked to lead the way to your vehicle.

Your vehicle will have been left unattended so walk around it and make a visual check of

- Lights
- Tyres
- Number plate
- Couplings (if appropriate)
- Cab locking mechanism (if fitted)

Before you start the engine you must always check that

- All doors are properly closed
- Your seat is properly adjusted and comfortable so that you can reach all the controls easily
- You have good all-round vision

Declaration

I declare that my use of the test vehicle for the purposes of the test is covered by a valid policy of insurance which satisfies the requirements of the relevant legislation

SIGNED

- Your driving mirrors are properly adjusted
- If fitted, your seat belt is fastened, correctly adjusted and comfortable
- The handbrake is on
- The gear lever is in neutral

Develop this routine while you're learning.

Once you've started the engine, if your vehicle is fitted with air brakes wait until the gauges show the correct pressure, or until any device (a buzzer sounding or a light flashing) has stopped operating.

As the driver of a large vehicle you must make safety your first priority – not just your own safety, but also that of other road users. The information in this book describes and encourages the safe driving techniques you should put into practice.

Driving in itself is a life skill, but when considering driving an LGV it's also the means by which you earn your living. It may take many years to gain the skills set out here, but you'll need to aim for professional standards right from the very start.

Whether you learn with an instructor or as 'work experience' with a colleague, you must be satisfied that you've fully covered all aspects of the officially recommended syllabus. You may find it helpful to check your progress against the actual syllabus requirements, so they're given here in full.

This syllabus lists the skills that you must have in order to reach the high standards required to pass the LGV driving test and to become a professional lorry driver. It's impossible to give details of all the rules and regulations that apply to both the driver and their vehicle in a book of this size. However, you'll need to know and keep up to date with current requirements.

Knowledge

You must have a thorough knowledge of

- The latest edition of *The Highway Code*, especially those sections that concern lorries
- Regulations governing drivers' permitted hours
- Regulations relating to the carriage of hazardous and other specialised goods

You must also have a thorough understanding of general motoring regulations, especially

- Road traffic offences
- Licences, both drivers' and operators', where applicable
- Insurance requirements
- Vehicle road tax relating to LGVs (and any trailer) in your charge
- Plating of LGVs and their trailers
- Annual testing of LGVs

Legal requirements

To learn to drive an LGV you must

1. Normally be at least 21 years old

2. Meet the stringent eyesight requirements

3. Be medically fit to drive lorries

4. Hold a full car licence (category B)

5. Hold and comply with the conditions for holding a provisional licence in the category of LGV being driven

6. Hold a full licence in category C and a provisional licence for C + E if driving a vehicle in that category

7. Ensure that the vehicle being driven

– is legally roadworthy

– is correctly plated

– has a current test certificate (MOT)

– is properly licensed with the correct tax disc displayed

8. Make sure that the vehicle being driven is properly insured for its use, especially if it's on contract hire

9. Display L plates to the front and rear of the vehicle (D plates, if you wish, when driving in Wales)

10. Be accompanied by a supervisor who holds a valid full licence for the category of vehicle being driven

11. Wear a seat belt, if fitted, unless you're exempt. Ensure that all seat belts in the vehicle, and their anchorages and fittings, are secure and free from obvious defects.

Children shouldn't normally be carried in LGVs. However, if a child is carried in the vehicle, with permission, you must comply with all regulations relating to the wearing of seat belts by children or the use of child restraints

12. Be aware that it's a legal requirement to notify the DVLA of any medical condition that could affect safe driving if the period is likely to be three months or more

13. Ensure that any adaptations are suitable to control the vehicle safely if the vehicle has been adapted for any disability

Vehicle controls, equipment and components

You must

1. Understand the function of the

- accelerator
- clutch
- gears
- footbrake
- handbrake
- secondary brake
- steering

and be able to use these competently

2. Know the function of all other controls and switches on the vehicle and be able to use them competently

3. Understand the meanings of

- gauges
- warning lights
- warning buzzers
- other displays on the instrument panel

4. Be familiar with the operation of tachographs and their charts

5. Know the legal requirements that apply to the vehicle

- speed limits
- weight limits
- braking system (ABS)
- fire extinguishers to be carried

6. Be able to carry out routine safety checks, and identify defects, especially on the

- power steering
- brakes (tractor unit + semi-trailer on articulated, or rigid vehicle + trailer on combinations)
- tyres on all wheels
- seat belts
- lights
- reflectors/reflective plates
- direction indicators
- marker lights
- windscreen, wipers and washers
- horn
- rear view mirrors
- speedometer
- tachograph
- exhaust system
- brake line and electrical connections on rigid vehicles + trailers, or articulated vehicles
- coupling gear
- hydraulic and lubricating systems
- self-loading or tailgate equipment
- drop-side hinges and tailgate fastenings

– curtain-side fittings/fastenings

– winches or auxiliary gear, where these items are fitted

7. Know the safety factors relating to

– stowage

– loading

– stability

– restraint

of any load carried on the vehicle

8. Know the effects speed limiters will have on the control of your vehicle, especially when you intend to overtake

9. Know the principles of the various systems of retarders that may be fitted to LGVs including

– electric

– engine-driven

– exhaust brakes

and when they should be brought into operation

Road user behaviour

You must

1. Know the most common causes of road traffic accidents

2. Know which road users are more vulnerable and how to reduce the risks to them

3. Know the rules, risks and effects of drinking and driving

4. Know the effects that

– illness (even minor ones)

– drugs or cold remedies

– tiredness

can have on driving performance

5. Recognise the importance of complying with rest period regulations

6. Be aware of the age-dependent problems among other road users including

– children

– young cyclists

– young drivers

– more elderly drivers

– elderly or infirm pedestrians

7. Concentrate and plan ahead in order to anticipate the likely actions of other road users and be able to select the safest course of action

Vehicle characteristics

You must

1. Know the most important principles concerning braking distances under various

- road
- weather
- loading

conditions

2. Know the different handling characteristics of other vehicles with regard to

- speed
- stability
- braking
- manoeuvrability

3. Know that some other vehicles, such as bicycles and motorcycles, are less easily seen than others

4. Be aware of the difficulties caused by the characteristics of both your own and other vehicles, and be able to take the appropriate action to reduce any risks that might arise.

Examples are

- LGVs and buses moving to the right before making a sharp left turn
- drivers of articulated vehicles having to take what appears to be an incorrect line before negotiating corners, roundabouts or entrances
- blind spots that occur on many large vehicles
- bicycles, motorcycles and high-sided vehicles being buffeted in strong winds, especially on exposed sections of road
- turbulence created by LGVs travelling at speed, affecting pedestrians, cyclists, motorcyclists, vehicles towing caravans, and drivers of smaller motor vehicles

At all times, remember that other road users may not understand the techniques required to manoeuvre an LGV safely

Road and weather conditions

You must

1. Know the various hazards that can arise when driving

- in strong sunlight
- at dusk or dawn
- during the hours of darkness
- on various types of road such as
  - country lanes in rural areas
  - one-way streets
  - two-way roads in built-up areas
  - three-lane roads
  - dual carriageways with various speed limits
  - trunk roads with two-way traffic
  - motorways

2. Gain experience in driving on urban roads with 20 or 30 mph speed limits, and also on roads carrying dense traffic volumes at higher speed limits in both daylight and during the hours of darkness

3. Gain experience in driving on both urban and rural motorways

4. Know which road surfaces will provide better or poorer grip when braking

5. Know all the associated hazards caused by bad weather such as

- rain
- snow
- ice
- fog

6. Be able to assess the difficulties caused by

- road
- traffic
- weather

conditions

7. Drive professionally and anticipate how the prevailing conditions may affect the standard of driving shown by other road users

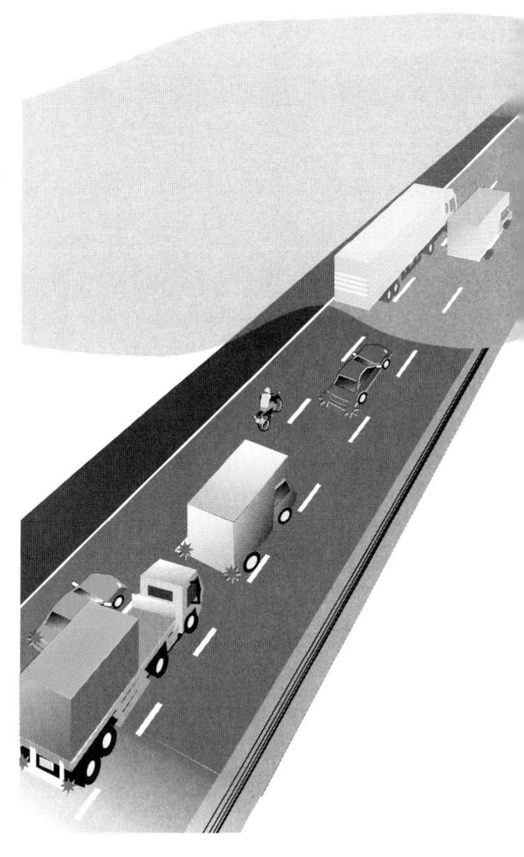

Traffic signs, rules and regulations

You must

1. Have a thorough knowledge and understanding of the meanings of traffic signs and road markings

2. You must be able to recognise and comply with traffic signs that indicate

- weight limits
- height limits
- when LGVs are prohibited
- loading/unloading restrictions
- traffic calming measures
  - 20 mph zones
  - road width restrictions
  - speed-reduction humps
- roads designated Red Routes
- night-time and weekend lorry bans such as those used in the London boroughs

Vehicle control and road procedure

You must have the knowledge and skill to carry out safely and expertly the following list of tasks (when appropriate) in daylight and, if necessary, during the hours of darkness. Where the tasks involve other road users you must

- Make proper use of the mirrors
- Take effective observation
- Give signals where necessary

1. Take the following necessary precautions, where they're applicable, before getting into the vehicle

- ensure that number plates are correct and securely fitted
- check all round for obstructions
- ensure that any load is secure
- check that air lines are correctly fitted and free from leaks
- check all couplings to the drawing vehicle and trailer
- check that the landing gear is raised
- check that the trailer brake is released
- check that all bulbs, lenses and reflectors are fitted

- make sure that all lights, indicators and stop lights are working
- ensure that all reflective plates are visible, clean and secure
- examine tyres for defects
- examine all load restraints for tension, etc.
- ensure that any unused ropes are safely stowed

2. Before leaving the vehicle cab make sure that

- the vehicle is stopped in a safe, legal and secure place
- the handbrake is on
- the engine is stopped
- the electrical system is switched off
- the gear lever/selector is in neutral
- all windows are closed
- the passenger door is secure
- the keys have been removed from the starter switch
- you won't endanger anyone when you open the door

3. Before starting the engine carry out the following safety checks

- handbrake is applied
- gear lever is in neutral
- doors are properly closed
- your seat is adjusted for
  - height
  - distance from the controls
  - back rest support and comfort
- The mirrors are correctly adjusted
- Your seat belt is fastened and adjusted

4. Start the engine, but before moving off check that

- the vehicle (and any trailer) lights are on, if required
- gauges indicate correct pressures for the braking system
- no warning lights are showing
- no warning buzzer is operating
- no ABS fault indicator is lit (where fitted)
- all fuel and temperature gauges are operating normally
- engine pre-heater lamp (glow plug) is operating (where fitted)
- it's safe to move off by looking all round, especially in the blind spots

5. Move off
– straight ahead
– at an angle
– on the level
– uphill
– downhill

6. Select the correct road position for normal driving

7. Practise effective observation in all traffic conditions

8. Drive at a speed appropriate to the road, traffic and weather conditions

9. Anticipate changes in traffic conditions and adopt the correct action at all times and exercise vehicle sympathy

10. Move into the appropriate traffic lane correctly and in good time

11. Pass stationary vehicles safely

12. Meet, overtake and cross the path of other vehicles safely

13. Turn right or left at
– junctions
– crossroads
– roundabouts

14. Drive ahead at crossroads and roundabouts

15. Keep a safe separation gap when following other vehicles

16. Act correctly at all types of pedestrian crossing

17. Show proper regard for the safety of all other road users, with particular respect for those most vulnerable

18. Drive on
– urban roads
– rural roads
– dual carriageways
keeping up with the traffic flow (but still observing speed limits) where it's safe and appropriate to do so

19. Comply with
– traffic regulations
– traffic signs
– signals given by authorised persons
  – police officers
  – traffic wardens
  – school crossing patrols

20. Take the correct action on signals given by other road users

21. Stop the vehicle safely at all times

22. Select safe and suitable places to stop the vehicle, when requested, reasonably close to the nearside kerb
– on the level
– facing uphill
– facing downhill
– before reaching a parked vehicle, but leaving sufficient room to move away again

23. Stop the vehicle on the braking exercise manoeuvring area

- safely
- as quickly as possible
- under full control
- within a reasonable distance from a designated point

24. Reverse the vehicle on the manoeuvring area

- under control
- with effective observation
- on a predetermined course
- to enter a restricted opening
- to stop with the extreme rear of the vehicle within a clearly defined area

25. Cross all types of level crossings

- railway
- rapid transit systems (trams)

where appropriate

26. Uncouple and recouple the tractor unit and trailer. When uncoupling you must

- select a place with safe and level ground
- apply the brakes on both the vehicle and trailer
- lower the landing gear
- stow the handle away safely
- turn off any taps fitted to the air lines
- disconnect the air lines and stow them away safely
- disconnect the electric lines and stow them away safely
- remove any 'dog clip' securing the kingpin release handle
- release the fifth wheel coupling locking bar, if fitted
- drive the tractor unit away slowly, checking the trailer either directly or in the mirrors
- take any anti-theft precautions (kingpin lock, etc.)
- remove the number plate

With a rigid vehicle and trailer you might have to support the trailer drawbar before you pull away.

When recoupling you must

- apply the trailer brake
- check that the height of the trailer is correct so that it will receive the unit safely
- reverse slowly up to the trailer ensuring that the kingpin locking mechanism is in place
- ensure that the locking mechanism is secure by selecting a low gear and attempting to move forward
- apply the parking brake before leaving the cab
- connect any 'dog clip' to secure the kingpin release handle
- connect the air and electric lines
- turn on taps, if fitted
- raise the landing gear and stow away the handle
- check that all electrics are working
- ensure that the trailer brake is released before moving off

- start up the engine and ensure that the gauges register correct pressures in air storage tanks and that no warning buzzer/light is operating
- obtain assistance to check for air line leaks and operation of all rear, marker or reversing lights, indicators, stop lights and fog light(s)
- secure the correct number plate and check that all reflectors are present and clean
- examine all tyres, wheel nuts, fastenings, ropes, sheets, drop-side locking clips, rear doors, hydraulic rams, tail-lift gear, etc. to ensure that the trailer and any load won't present a danger to other road users
- check the function of any ABS warning lights, etc.
- make sure that your mirrors are properly adjusted to give the best view down each side of the trailer before driving off
- test the operation of the brakes at a safe place, ideally before moving out onto a public road

When recoupling a rigid vehicle + trailer combination the sequence is similar, but the trailer drawbar will need to be adjusted to the correct height (often by means of a 'bottle' jack) before the towing vehicle reverses to recouple the trailer. Be on the alert for the safety of anyone at the rear of your vehicle who is assisting you to recouple the trailer.

L

This part looks at what the practical driving test requires.

The topics covered

- The reversing exercise
- The braking exercise
- The vehicle controls
- Other controls
- The gear-changing exercise
- Moving off
- Using the mirrors
- Giving signals
- Acting on signs and signals
- Awareness and anticipation
- Making progress
- Controlling your speed
- Separation distance
- Hazards
- Selecting a safe place to stop
- Uncoupling and recoupling
- If you pass
- If you don't pass

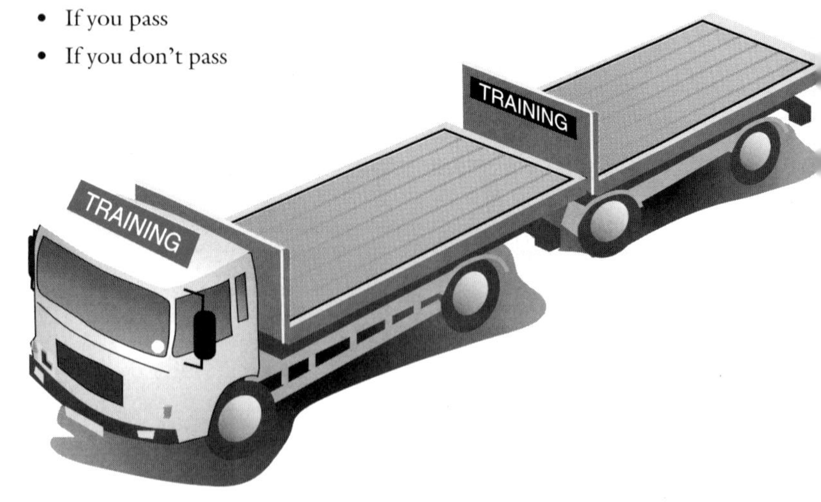

The exercise is started from a position with the front of the vehicle in line with cones A and A1. You should reverse your vehicle into the bay, keeping marker B on the offside. You should stop with the extreme rear of your vehicle within the 90 cm stopping area.

The distances

A to A1 = 1.5 times the width of the vehicle

A to B = twice the length of the vehicle

B to line Z = 3 times the length of the vehicle

The length of the bay will be based on the length of the vehicle. This can be varied at the discretion of the examiner within the range of 1 and 2 metres (3 and 6 feet).

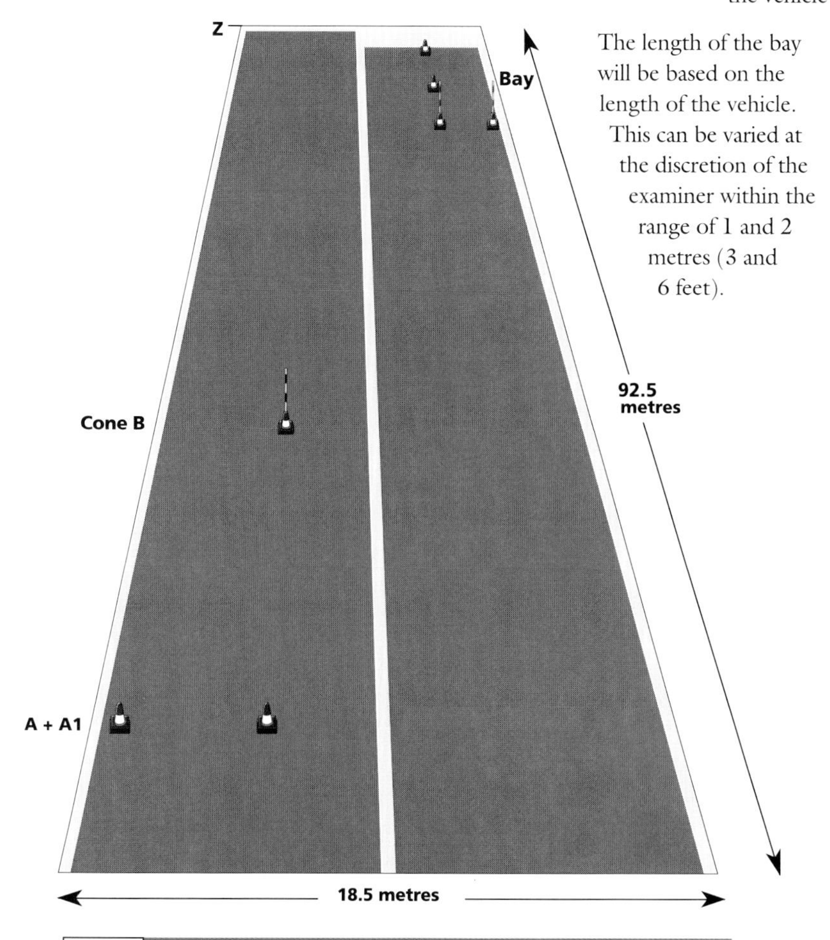

What the test requires

You should be able to reverse your vehicle and trailer in a restricted space. You must be able to do this

- Under control and in reasonable time
- With good observation
- With reasonable accuracy
- Starting at a fixed point (cones A and A1)
- Inside a clearly defined boundary (yellow lines)
- By reversing so that you pass cone B on the offside of your vehicle
- So that the extreme rear of your trailer is stopped within the painted yellow box area at the end of the bay

Your examiner will show you a diagram of the manoeuvring area and explain what's required.

Skills you should show

You should complete the exercise

- Reversing under complete control
- Using good, effective, all-round observation
- Ensuring accurate judgement of the size of your vehicle and trailer from the cab
- Driving with careful co-ordination of the clutch, accelerator and footbrake until the exercise is completed

Faults you should avoid

- Approaching the starting point too fast
- Not driving in a straight line as you approach cones A and A1
- Stopping beyond the first marker cones A and A1
- Turning the steering wheel incorrectly when starting to reverse
- Over-steering so that any wheel goes over the yellow boundary
- Not practising effective observation or misjudging the position of your vehicle so that it comes in contact with any cone or pole
- Incorrect judgement so that the rear of your vehicle and trailer is either short of or beyond the yellow box in the bay
- Using excessive steering movements or shunts to complete the manoeuvre
- When taking a forward shunt, driving down the area ahead of a position level with cones A and A1
- Carrying out the manoeuvre at a very slow pace

There's no emergency stop exercise in the LGV driving test. For safety reasons a braking exercise takes place at a special area and not on the public roads.

What the test requires

Your examiner will be with you in the vehicle for this exercise.

She or he will point out two marker cones about 61 metres (200 feet) ahead. You should build up a speed of about 20 mph. When the front of the vehicle passes between the two markers you should apply the brakes.

You must stop your vehicle and trailer with safety and under full control.

Skills you should show

You should stop the vehicle

- As quickly as possible
- Under full control
- As safely as possible
- In a straight line

Faults you should avoid

- Driving too slowly (less than 20 mph)
- Braking too soon (anticipating the marker points)
- Braking too harshly, causing skidding
- Depressing the clutch too early, losing engine braking and possibly causing the wheels to lock
- Depressing the clutch too late, stalling the engine
- Taking too long to stop

What the test requires

You must show your examiner that you understand the functions of all the controls. You should use them

- Smoothly
- Correctly
- Safely
- At the right time

The main controls are

- Accelerator
- Clutch
- Footbrake
- Handbrake
- Steering
- Gears

You must

- Understand what these controls do
- Be able to use them competently

How your examiner will test you

For this aspect of driving there isn't a special exercise. Your examiner will watch you carefully to see how you use these controls.

Skills you should show

Accelerator and clutch

- Balance the accelerator and clutch to pull away smoothly
- Accelerate gradually to gain speed
- When stopping the vehicle, press the clutch in just before it stops

If your vehicle has automatic transmission ensure that your foot is on the footbrake when you engage 'drive' (D).

Gears

The gears are designed to assist the engine to deliver power under a variety of conditions. The lowest gears may only be necessary if the vehicle is loaded or when climbing steep gradients. The gearbox may have one or more 'crawler' gear positions.

You should

- Move off in the most suitable gear
- Change gear in good time before a junction or hazard
- Show an understanding of the type of gearbox you're using by demonstrating its abilities
- Plan well ahead, whether climbing or before starting to descend a long hill

If you leave it until you're either losing or gaining too much speed you may have difficulty selecting gears and maintaining control.

Modern vehicles may be fitted with sophisticated systems controlled by computer. These systems can sense the load, speed or gradient and select the correct gear for the conditions. With these systems the driver may only have to ease the accelerator, or depress the clutch pedal, to allow the system to engage the gear required.

If you're driving a vehicle with a 'range change' gearbox you should ensure that you've selected the correct range before changing gear.

Faults you should avoid

Accelerator

- Loud over-revving, causing excessive engine noise and exhaust fumes, and also alarming or distracting other road users

Clutch

- Jerky and uncontrolled use of the clutch when moving off or changing gear

Gears

- Taking your eyes off the road when you change gear
- 'Coasting' with the clutch pedal depressed or leaving the gear lever in neutral. This is highly dangerous if you're driving a vehicle with air brakes. The engine-driven compressor won't be able to replace the air being used by the brakes due to the engine running only at idle speed
- Holding onto the gear lever unnecessarily
- Forgetting to move the range selector switch

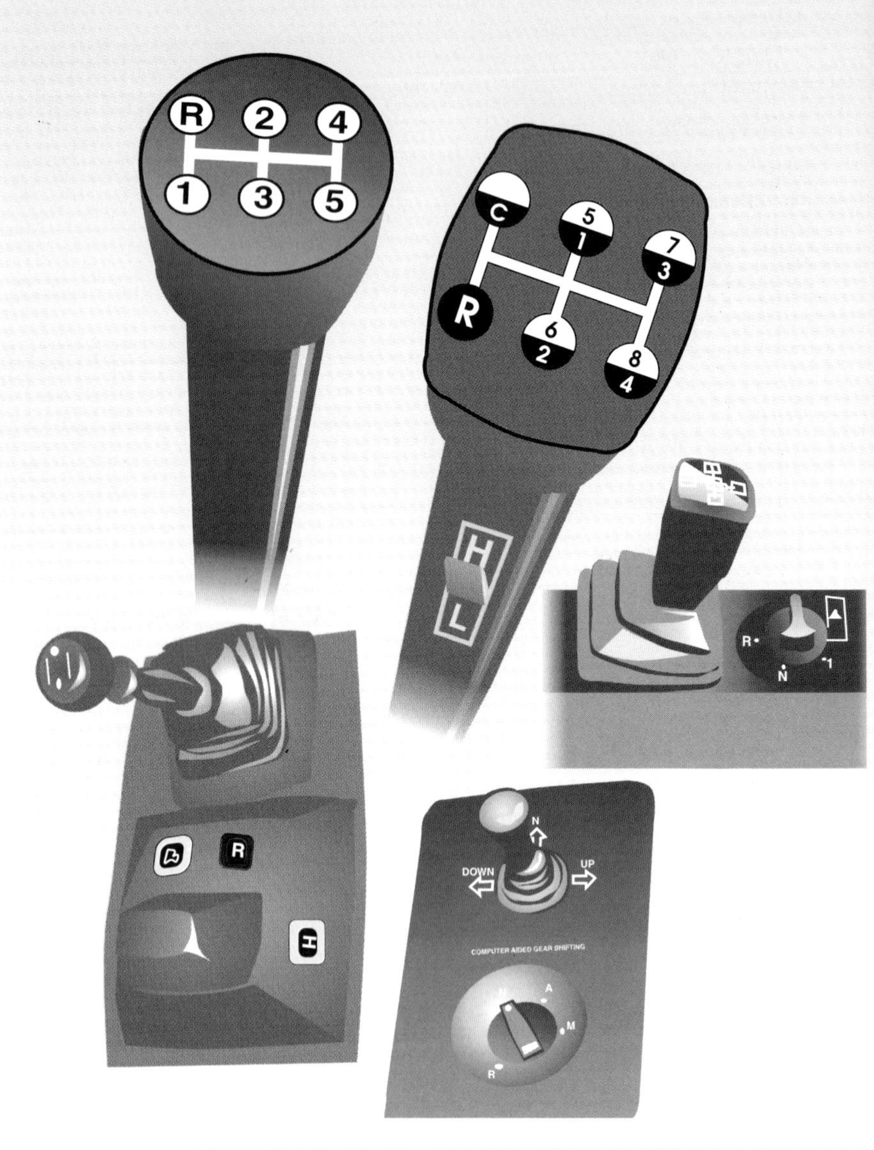
R
2
4
1
3
5
C
5
1
7
3
R
6
2
8
4
H
L
R
N
1
R
N
DOWN
UP
COMPUTER AIDED GEAR SHIFTING
N
A
M
R

Skills you should show

Brake

- In good time
- Lightly, in most situations
- Progressively

Most large vehicles are equipped with air brake systems. There's no direct relationship between the pressure applied to the pedal and the braking force exerted on the wheels. This means that good control is needed at all times.

Faults you should avoid

- Braking harshly
- Excessive and prolonged use of the footbrake
- Braking and steering at the same time, unless already travelling at a low speed

Air brake system

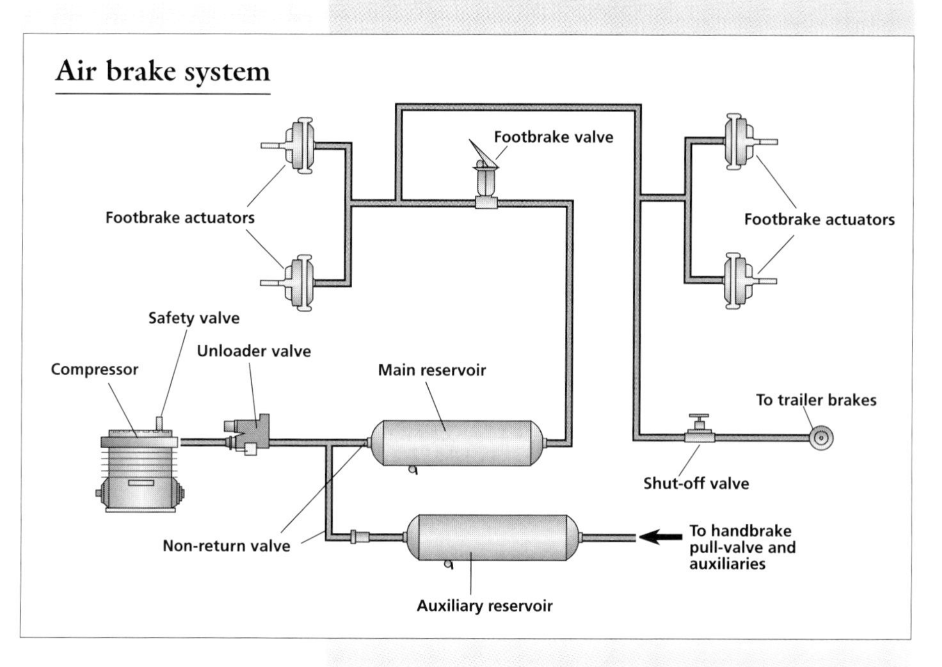

Skills you should show

Handbrake

You should know how and when to apply the handbrake effectively.

Some modern vehicles will apply a parking brake when the vehicle is brought to a stop by the footbrake.

Steering

You should

- Place your hands on the steering wheel in a position that's comfortable and which gives you full control
- Keep your movements steady and smooth
- Turn the steering wheel to turn a corner at the correct time

Power-assisted steering

Most modern vehicles are fitted with power-assisted steering. This relieves the driver of steering effort, especially at slow speeds.

When the engine is running, hydraulic pressure is built up in the system to make the steering easier. If the steering becomes 'heavy' check for leaks in the system.

Faults you should avoid

Handbrake

- Applying the handbrake before the vehicle has stopped
- Trying to move off with the handbrake on
- Allowing the vehicle to roll back as you move off

Steering

Don't steer too early when turning a corner. If you do, you risk

- Cutting the corner when turning right, causing the rear wheels to cut across the path of traffic waiting to emerge
- Striking the kerb when turning left

Don't turn too late. You could put other road users at risk by

- Swinging wide at left turns
- Overshooting right turns

Avoid

- Crossing your hands on the steering wheel whenever possible
- Allowing the wheel to spin back after turning
- Resting your arm on the door

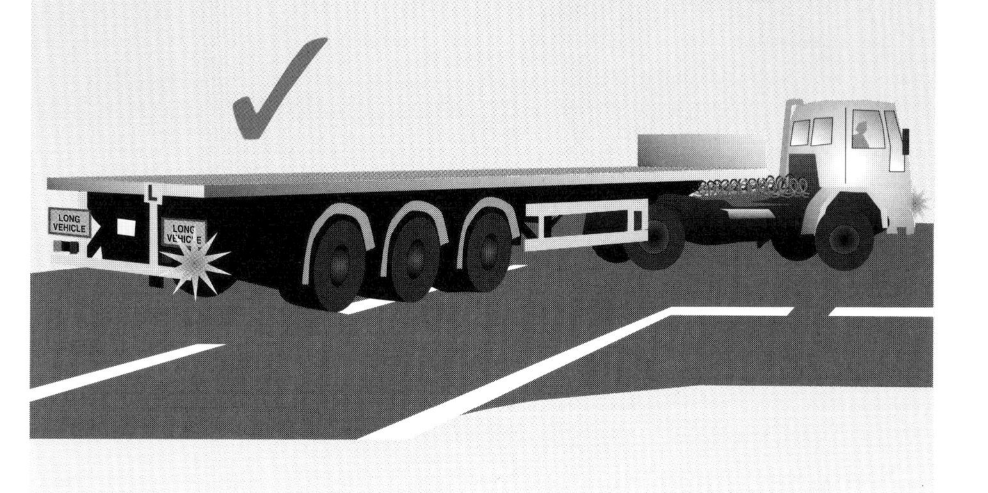
LONG
VEHICLE
L

LONG
VEHICLE
L

You should understand

The functions of all controls and switches on your vehicle that have a bearing on road safety, such as

- Indicators
- Lights
- Windscreen wipers
- Demisters

You should know the meaning of all gauges and switches on the instrument panel, such as

- Air pressure gauges
- Speedometer
- Various warning lights and buzzers
- On-board computer displays
- ABS failure warnings
- Bulb failure warnings
- Gear-selection indicators

Safety checks and fault recognition

You should be able to carry out routine checks, such as on

- Steering
- Brakes
- Tyres
- Seat belts
- Lights
- Reflectors
- Horn
- Rear view mirrors
- Speedometer
- Exhaust system
- Direction indicators
- Windscreen, wipers and washers
- Wheel-nut security

You should be able to understand the effect any fault may have on your vehicle.

What the test requires

Your examiner wants to see you engage the lower gears competently. These gears may not otherwise be used during the drive.

You'll be asked to pull up at a convenient place to carry out the gear-changing exercise. Your examiner will then ask you to move off in your *lowest* gear and to change up to each gear in turn. If your vehicle has a 'crawler' gear, which requires the vehicle to be stationary when engaged, you won't be asked to use it during the gear-changing exercise.

Drive for a short distance in each gear until you reach a gear your examiner considers appropriate for your vehicle. This will depend on the gearbox layout.

When you've reached the highest gear your examiner will ask you to change back down into each gear in turn until you reach the lowest gear again.

Skills you should show

- A smooth start in the lowest gear
- Judging the speed correctly for the next gear
- Smooth engagement of the next lowest gear by bringing the speed of the vehicle down and by careful use of the footbrake, if necessary
- Returning to the lowest gear without stopping the vehicle completely

Throughout the exercise you should

- Practise effective observation, especially before moving off and slowing down
- Give any signal that might be appropriate
- Use the controls skilfully to ensure smooth engagement of the gears

Faults you should avoid

- Not checking the blind spots before moving off or slowing down, and not acting on what you see in the mirrors
- Jerky use of the clutch or accelerator
- Not starting off in the lowest gear
- Not selecting the next gear in the sequence
- Not being able to engage a gear
- Not giving a signal to any following traffic before slowing down

What the test requires

You should be able to move off safely and under control

- On the level
- From behind a parked vehicle
- Uphill
- Downhill

How your examiner will test you

Your examiner will watch your use of the controls as you move off.

Skills you should show

Before you move off

- Use your mirrors
- Look all around your vehicle

You should be aware of

- Other vehicles
- Cyclists
- Pedestrians outside the range of your mirrors

You should move off under control, making balanced and safe use of the

- Accelerator
- Clutch
- Brakes
- Steering
- Correct gear

Uphill

- Practise effective observation before moving off, checking all blind spots
- Give a signal, if required, at the correct time
- Use sufficient acceleration, depending on the gradient
- Move off smoothly
- Change up as soon as it's safe to do so

Downhill

- Move off only when it's safe to do so
- Practise effective observation before moving off, checking all blind spots
- Give a signal, if required, at the correct time

When you move off

- Engage the correct gear for the gradient
- Hold the vehicle on the footbrake and release the handbrake until it's safe to move away
- Co-ordinate the clutch and accelerator
- Build up speed when it's safe to do so

Faults you should avoid

Uphill

Moving off without

- Using both nearside and offside mirrors
- Looking around to check all blind spots
- Giving a correct signal if it's required
- Using enough revs for the gradient

You should avoid moving off without good control of the accelerator, clutch and handbrake. Don't

- Stall
- Roll backwards
- Surge away

Downhill

Moving off without

- Using both nearside and offside mirrors
- Looking around to check all the blind spots
- Giving a correct signal, if it's required
- Co-ordinating the accelerator, clutch and handbrake so that the vehicle stalls or surges

How your examiner will test you

At an angle

Your examiner will ask you to pull up on the left just before you reach a parked vehicle. You'll be asked to move away to show your ability to move off at an angle.

Skills you should show

When moving out from behind a parked vehicle you should

- Practise effective all-round observation
- Check any blind spots
- Give a signal, if it's necessary
- Move out only when it's safe to do so
- Move out well clear of the parked vehicle
- Check your mirrors, especially the nearside, to confirm that you're clear of the parked vehicle

Faults you should avoid

- Pulling out unsafely
- Causing other road users to stop or alter their course
- Excessive acceleration
- Moving off in too high a gear
- Failing to co-ordinate the controls correctly and stalling the engine
- Swinging excessively wide into the path of oncoming traffic

What the test requires

You must always use your mirrors effectively

- Before any manoeuvre
- To keep up to date on what's happening behind you

Use them before

- Moving off
- Signalling
- Changing direction
- Turning left or right
- Overtaking or changing lanes
- Increasing speed
- Slowing down or stopping
- Opening your cab door

Check again in your nearside mirror after passing

- Parked vehicles
- Horse riders, motorcyclists or cyclists
- Any pedestrians standing close to the kerb
- Any vehicle that you've just overtaken

before moving back into the left

How your examiner will test you

For this aspect of driving there isn't a special exercise. Your examiner will watch you use your mirrors as you drive.

Skills you should show

Use the Mirrors – Signal – Manoeuvre (MSM) routine and also the Position – Speed – Look (PSL) routine.

- Look before you signal
- Signal before you act
- Act sensibly on what you see in the mirrors
- Be aware that the mirrors won't show everything behind you

You should check your nearside mirror every time you pass

- Parked vehicles
- Vulnerable road users
- Vehicles you've just overtaken
- Pedestrians near the kerb

Always have as good an idea of what's happening behind you as what's going on in front. You should also be aware of the effect your large vehicle has on other road users around you.

What the test requires

You must give clear signals in good time so that other road users know what you're about to do. This is particularly important with LGVs because other road users may not understand the position you need to move into

- Before turning left
- Before turning right
- At roundabouts
- To move off at an angle
- Before reversing into an opening

You should only give signals that are in *The Highway Code.*

Correct signals should help other road users to

- Understand what you intend to do next
- Take appropriate action

Always check that you've cancelled an indicator as soon as it's safe to do so.

How your examiner will test you

For this aspect of driving there isn't a special exercise. Your examiner will watch you carefully to see how you use your signals as you drive.

Skills you should show

You should give signals

- Clearly
- At the appropriate time
- By indicator
- By arm, if necessary

Faults you should avoid

- Giving misleading or incorrect signals
- Omitting to cancel signals
- Waving on pedestrians to cross in front of your vehicle
- Giving signals other than those shown in *The Highway Code*

What the test requires

You should have a thorough knowledge of traffic signs, signals and road markings. You should be able to

- Recognise them in good time
- Take appropriate action on them

At the start of the drive your examiner will ask you to follow the road ahead, unless traffic signs indicate otherwise or you're asked to turn. From this point your examiner will expect you to understand and act correctly on road signs or signals that occur.

Skills you should show

Traffic lights and signals

You should

- Comply with traffic lights and signals
- Approach at a speed that allows you to stop, if necessary, under full control
- Only move forward at a green traffic light if it's clear for you to do so and you won't block the junction

Authorised persons

You must comply with the signals given by

- Police officers
- Traffic wardens
- School crossing patrols
- An authorised person controlling the traffic, e.g., at road repairs

Other road users

You should look out for signals given by other road users and

- React safely
- Take appropriate action
- Anticipate their action
- Use your brakes and/or give arm signals, if necessary, to any traffic following your vehicle

Those behind you may not be able to see signals given by a road user ahead of you. Giving them a signal is particularly important if a driver or rider ahead of you is intending to turn and the size of your vehicle prevents traffic behind you from seeing their signal.

What the test requires

You must be aware of other road users at all times. You should also always plan ahead.

- Judge what other road users are going to do
- Predict how their actions would affect you
- React safely and in good time

Skills you should show

- Awareness of and consideration for all other road users
- Anticipation of possible danger and concern for safety

Pedestrians

- Give way to pedestrians when turning from one road into another
- Take particular care with the very young, the disabled and the elderly. They may not have seen you and might not be able to react quickly to danger

Cyclists

Take special care

- When crossing bus or cycle lanes
- With cyclists passing on your left
- With child cyclists

Moped riders and motorcyclists

Look out for mopeds and motorcyclists

- In slow-moving traffic
- Coming up on your left
- At junctions

Horse riders and animals

Take special care with people in charge of animals, especially horse riders.

Faults you should avoid

- Reacting suddenly to road or traffic conditions rather than anticipating them
- Showing irritation with other road users
- Sounding the horn aggressively
- Revving your engine or edging forward when waiting for pedestrians to cross

Your examiner will be looking for a high standard of driving from an experienced driver. You'll need to display safe, confident, positive driving techniques.

How your examiner will test you

For this aspect of driving there isn't a special exercise. As an experienced driver you'll be expected to drive accordingly. Your examiner will watch your driving and expect you to make good progress safely.

Skills you should show

You should

- Make reasonable progress where conditions allow
- Keep up with the traffic flow when it's safe to do so
- Make positive, safe decisions as you make progress

You should be able to drive at the appropriate speed depending on the

- Type of road
- Weather conditions and visibility
- Traffic conditions

Approach all hazards at a safe speed without

- Being unduly cautious
- Holding up following traffic unnecessarily

Faults you should avoid

- Driving so slowly that you hinder other traffic
- Being over-cautious or hesitant
- Stopping when you can see it's obviously clear and safe to go on

What the test requires

You should make good progress along the road, taking into consideration

- The type of road
- The volume of traffic
- The weather conditions and the state of the road surface
- The braking characteristics of your vehicle
- Speed limits that apply to your vehicle
- Any hazards associated with the time of day (schools, etc.)

How your examiner will test you

For this aspect of driving there isn't a special exercise. Your examiner will watch you control your speed as you drive.

Skills you should show

- Take great care in the use of speed
- Drive at the correct speed for the traffic conditions
- Be sure that you can stop safely in the distance that you can see to be clear
- Leave a separation distance between your vehicle and the traffic ahead
- Allow extra stopping distance on a wet or slippery surface
- Observe the speed limit that applies to your vehicle
- Anticipate any hazards that could arise
- Allow for the mistakes of others

Faults you should avoid

- Driving too fast for the conditions
- Exceeding speed limits
- Varying your speed erratically
- Having to brake hard to avoid a situation ahead
- Approaching bends, traffic signals and any other hazards too fast

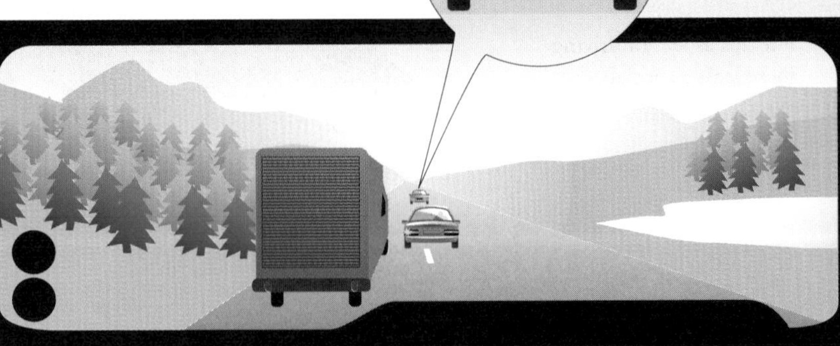

Always keep a safe separation distance between you and the traffic in front.

What the test requires

You should always drive so that you can stop safely in the distance that you can see to be clear.

In good weather conditions, leave a gap of at least 1 metre (about 3 feet) for each mph of your speed, or a two-second time gap.

In bad conditions, leave at least double that distance, or a four-second time gap.

In slow-moving congested traffic it may not be practical to leave as much space.

How your examiner will test you

For this aspect of driving there isn't a special exercise. Your examiner will watch you as you drive and take account of your

- Use of the MSM/PSL routine
- Anticipation
- Reaction to changing road and traffic conditions
- Handling of the controls

Skills you should show

- Judge a safe separation distance from the traffic in front
- Show correct use of the MSM/PSL routine, especially before reducing speed
- Avoid the need to brake sharply if the vehicle in front slows down or stops
- Take extra care when your view ahead is limited by large vehicles
- Keep a good separation distance from traffic queues in front

Look out for

- Brake lights ahead
- Direction indicators
- Vehicles ahead braking without warning

Faults you should avoid

- Following too closely or 'tailgating'
- Braking suddenly
- Swerving to avoid the vehicle in front, which may be slowing down or stopping

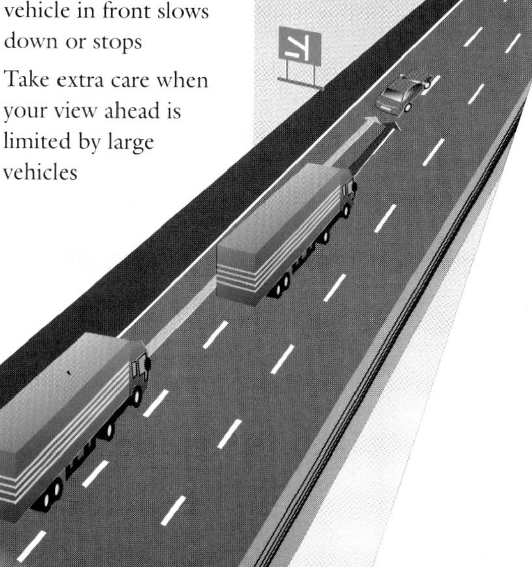

HAZARDS – THE CORRECT ROUTINE

A hazard may present itself when you're either stationary or on the move. Hazards may be included in any situation that involves you adjusting your speed or altering your course. In addition, hazards can be created by the actions of other road users around you. Look well ahead for

- Road junctions or roundabouts
- Parked vehicles
- Cyclists or horse riders
- Pedestrian crossings
- Pedestrians on or near the kerbside
- Cyclists or motorcyclists moving up alongside
- Drivers edging up on the nearside before you make a turn
- Vehicles pulling up close behind when you want to reverse

Watch what's happening around you.

Traffic situations are constantly changing. These changes might depend on the

- Time of day
- Location
- Density of traffic

You should be aware and anticipate potential hazards when you drive.

Skills you should show

Your higher seat position in an LGV may mean that you're able to see more clearly some of the hazards around you. As a professional driver you must be able to anticipate what might happen. You should be driving with a sense of awareness and anticipation. Know

- What's happening ahead
- What other road users are about to do
- When to take action

Scan the road ahead and be alert in case you have to

- Speed up
- Slow down
- Prepare to stop
- Change direction

Skills you should show

Pedestrians

Give way to pedestrians when turning from one road into another or when entering premises such as supermarkets, shops, warehouses, etc. Take extra care with the

- Young
- Elderly

and people who appear to have a disability.

Look out for pedestrians at all times but especially in shopping areas, where there might be a number of people waiting to cross the road, often at junctions.

Drive slowly and considerately when you need to enter pedestrianised areas to deliver to premises, during times when unloading or loading is permitted.

Cyclists

Take extra care when

- Crossing cycle lanes
- You're about to turn left and you can see a cyclist near the rear of your vehicle or moving up along the nearside of your vehicle
- Approaching any children on cycles
- There are gusty wind conditions

Motorcyclists

Look out for motorcyclists who are

- 'Filtering' in slow traffic streams
- Moving up alongside your vehicle, especially the nearside

Be especially aware when you're waiting to move out from a junction.

- Think once
- Think twice
- **Think bike**

Horse riders and animals

The size and noise of your vehicle can easily unsettle a horse.

- Give riders as much room as is safe
- Avoid revving the engine
- Look out for young riders who are learning and might not be able to control their horses
- React in good time to anyone who is herding animals
- Look out for warning signs, e.g., cattle

Faults you should avoid

- Sounding the horn aggressively
- Revving the engine and causing the air brakes to 'hiss'
- Edging forward when pedestrians are crossing in front
- Showing any sign of irritation

What the test requires

You should

- Normally keep well to the left
- Keep clear of parked vehicles
- Avoid weaving in and out between parked vehicles
- Position your vehicle correctly for the direction you intend to take

You should obey all lane markings, especially

- Left- or right-turn arrows at junctions
- When approaching roundabouts
- In one-way streets
- For bus lanes
- Road markings for LGVs approaching arches or narrow bridges with restricted headroom

How your examiner will test you

For this aspect of driving there isn't a special exercise. Your examiner will watch carefully to see that you

- Use the MSM routine
- Select the correct lane in good time

Skills you should show

You should

- Plan ahead and choose the correct lane in good time
- Use the MSM routine correctly
- Position your vehicle sensibly, even if there aren't any road markings
- Be aware that other road users might not understand your actions, so signal in good time

Faults you should avoid

- Driving too close to the kerb
- Driving too close to the centre of the road
- Changing lanes at the last moment or without good reason
- Hindering other road users by being badly positioned or in the wrong lane
- Straddling lanes or lane markings
- Cutting across the path of other traffic in another lane at roundabouts

There may be occasions, due to the length of your vehicle, when you have to straddle lane markings to avoid mounting the kerb or colliding with lamp-posts, traffic signs, etc. Use your own skill and judgement to make these decisions.

What the test requires

The size of your vehicle means that it's essential to make the correct decisions at junctions. Look well ahead and assess the situation as you approach.

Judge carefully when it's safe to emerge. Act on what you see. Wait until you can see that it's clear, then move away safely.

If you don't know, don't go.

Skills you should show

You should

- Use the MSM/PSL routine in good time when you approach a junction or a roundabout
- Assess the situation correctly so that you can position your vehicle to negotiate the junction safely
- Take as much room as you need as you approach a junction, if you're driving a long rigid vehicle or a vehicle and trailer combination
- Be aware of any lane markings and the fact that your vehicle may have to occupy part of the lane alongside
- Position as early as it's practicable to do so in one-way streets
- Make sure that you take **effective** observation before emerging into any junction

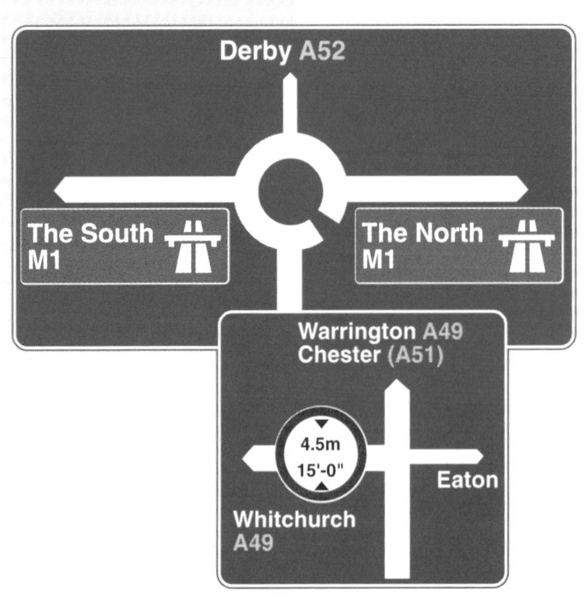

Left or right turns

Extra care should be taken if you're driving a long or an articulated vehicle, or a vehicle and trailer combination.

When turning left or right, position your vehicle so that you can

- See into the road
- Turn without the rear/trailer wheels mounting the kerb
- Turn without any part of the vehicle colliding with bollards or guard rails

If you're crossing a dual carriageway or turning right onto one, don't move forward unless you can clear the centre reservation safely. If your vehicle is too long for the gap, wait until it's clear from both sides and there's a safe opportunity to go.

Always

- Use your mirrors to check the rear wheels of your vehicle or trailer as you turn into or out of a junction
- Assess the speed of oncoming traffic correctly before crossing or entering roads with fast-moving traffic
- Allow for the fact that you'll need more time to build up speed in the new road

Observe, assess, then judge before you act.

What the test requires

Roundabouts can vary in size and complexity but the object of them all is to allow traffic to flow wherever possible.

Some roundabouts are so complex or busy that they require traffic lights to control the volume of traffic. The lights may be used at peak times when the traffic becomes very heavy.

At the majority of roundabouts the approaching traffic is required to give way to traffic approaching from the right. There are some locations where a 'give way' sign and markings apply to traffic already on the roundabout. You must be aware of these differences.

Skills you should show

It's essential that you plan your approach well in advance and use the MSM/PSL routine in good time. It's most important that you get into the correct lane as you approach. You should know the exit you wish to take and choose the suitable lane, taking into consideration the size of your vehicle.

On approach

You should adjust the speed of your vehicle in good time on the approach to roundabouts. Adjust your speed and select the correct gear to negotiate the roundabout safely. Get into the correct lane.

- Plan well ahead
- Look out for traffic signs as you approach
- Have a clear picture of the exit you intend to take
- Look out for the number of exits before yours
- Either follow the lane markings, as far as possible, or select the lane most suitable for the size of your vehicle
- Use the MSM/PSL routine in good time
- Signal your intentions in good time
- Avoid driving into the roundabout too close to the right-hand kerb. If you're driving an articulated vehicle you'll have to steer to the left to avoid the roundabout kerb

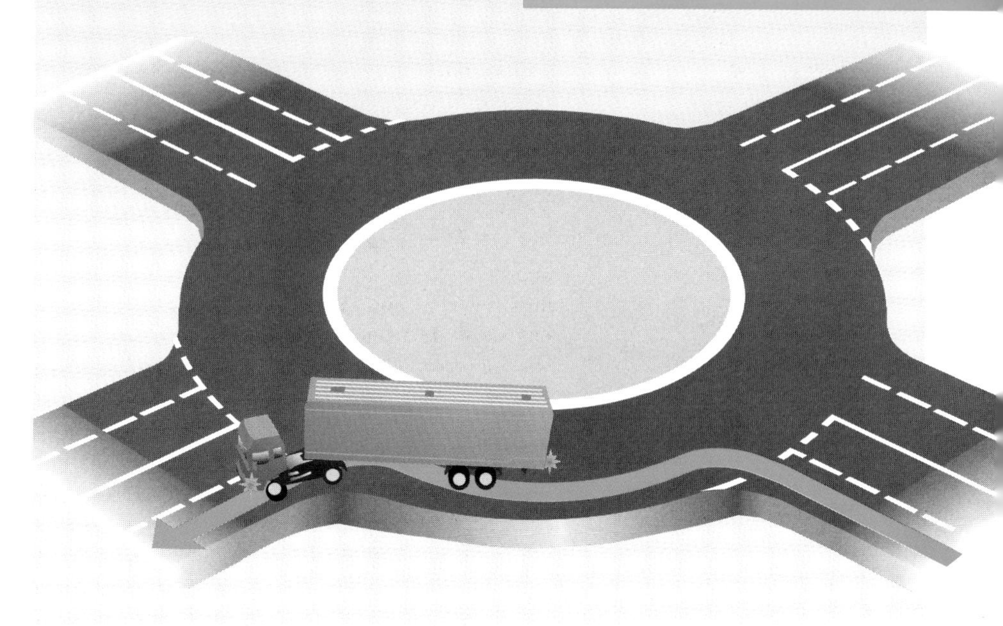

Whenever you enter a roundabout watch the vehicle in front of you. Make sure that it hasn't stopped while you were looking to the right. The driver ahead might be hesitant, so don't drive into the rear of the vehicle in front.

Turning left

- Give a left-turn signal in good time as you approach
- Approach in the left-hand lane. If you're driving a long vehicle you might need to take some of the lane on your right
- Adopt a path that ensures that the rear/trailer wheels don't mount the kerb
- Give way to traffic approaching from the right
- Use the nearside mirrors to be sure that no cyclists or motorcyclists are trapped along the nearside
- Continue to signal through the turn
- Look well ahead for traffic islands or bollards in the centre of your exit road. These could restrict the width available to you

Going ahead

(Up to 12 on a clock face)

- Approach in the left-hand lane unless blocked or clearly marked for 'left turn' only
- Don't give a signal on approach
- Try to stay in the lane, depending on the length of your vehicle
- Keep checking the mirrors both nearside and offside
- Give way to traffic from the right, if necessary
- Indicate left as you pass the exit just before the one that you intend to take
- Look well ahead for traffic islands or bollards in the centre of your exit road
- Make sure that the rear/trailer wheels don't mount the kerb as you leave the roundabout

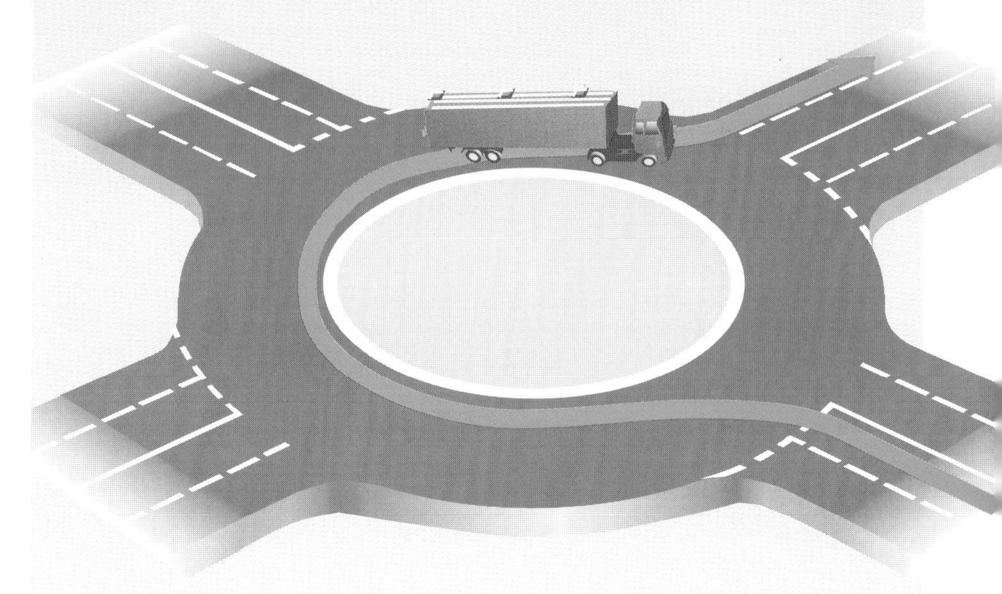

Turning right or full circle

- Look well ahead and use the MSM/PSL routine
- With a long vehicle, if there's a choice of two lanes for turning right use the left-hand of the two lanes
- If only one lane is marked for right turns you might have to occupy part of the lane on your left, not only on approach but also through the roundabout
- Signal right in good time before moving over to the right on approach
- Look out for any traffic accelerating up on the offside of your vehicle

Mini-roundabouts

- Give way to the traffic from the right
- You might have restricted room, so keep a constant check in the mirrors
- Position your vehicle correctly on approach so that you don't mount any kerbs
- Understand that other road users might not be aware of the room you need to complete a turn

Multiple roundabouts

Plan your approach early. You should

- Ensure that your eventual exit is clear so that you don't block the roundabout
- Give clear signals
- Look out for road users carrying out U-turns

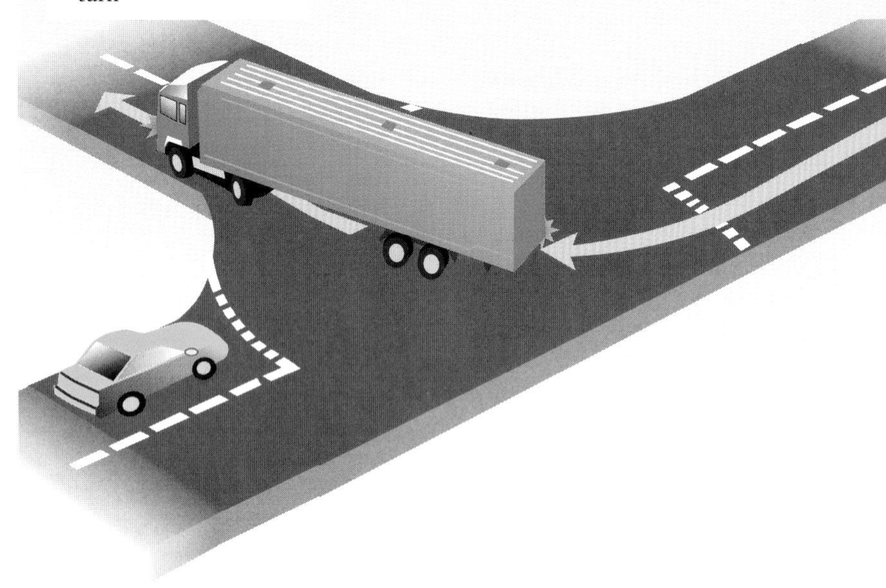

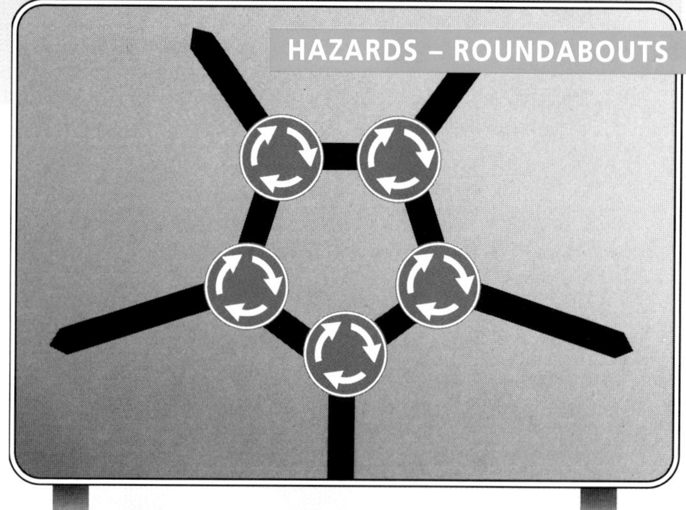

In a number of locations complex roundabout systems have been designed, which incorporate a mini-roundabout at each exit. The main thing to remember at such places is that traffic will be travelling in all directions. You must give way to traffic on the right.

Road surfaces

At the entrances to roundabouts there will have been a great deal of braking and accelerating. This will often make the road surface slippery, especially when it's wet. You should

- Brake in good time
- Enter the roundabout ensuring that other road users don't have to brake suddenly or swerve

Cyclists and horse riders

These road users might be in the left-hand lane on approach but may intend to turn right. Be aware of this and give them plenty of room.

What the test requires

Before overtaking you should look well ahead for any hazards, such as

- Oncoming traffic
- Bends
- Junctions
- The vehicle in front about to overtake
- Any gradient
- Those pointed out by road markings or traffic signs

You should assess the

- Speed of the vehicle that you intend to overtake
- Speed differential of the two vehicles, to judge how long the manoeuvre could take
- Time you have to complete overtaking safely

Avoid the need to 'cut in' on the vehicle that you've just overtaken.

How your examiner will test you

For this aspect of driving there isn't a special exercise. Your examiner will watch you and take account of your

- Use of the MSM/PSL routine
- Reaction to road and traffic conditions
- Handling of the controls
- Choice of a safe opportunity to overtake

Skills you should show

You should be able to assess all the factors that will decide if you can or can't overtake safely, such as

- Oncoming traffic
- The type of road
- The speed of the vehicle ahead
- Continuous white line markings on your side of the road
- How far ahead the road is clear
- Whether the road will remain clear
- Whether traffic behind is about to overtake your vehicle

Only overtake where you can do so safely, legally and without causing other road users to slow down or alter course.

Faults you should avoid

Don't overtake when

- Your view of the road ahead isn't clear
- You'd have to exceed the speed limit
- To do so would cause other road users to slow down or stop
- There are signs or road markings that prohibit overtaking

What the test requires

You should be able to meet and deal with oncoming traffic safely and confidently

- On narrow roads
- Where there are obstructions such as parked cars
- Where you have to move into the path of oncoming traffic

How your examiner will test you

For this aspect of driving there isn't a special exercise. Your examiner will watch you and take account of your

- Use of the MSM/PSL routine
- Reactions to road and traffic conditions
- Handling of the controls

Skills you should show

You should

- Show good judgement when meeting other traffic
- Be decisive when stopping and moving off
- Stop in a position that allows you to move out smoothly when the way is clear
- Allow adequate clearance when passing stationary vehicles. Slow right down if you have to pass close to them

Look out for

- Doors opening
- Children running out
- Pedestrians stepping out from between parked cars or from buses
- Vehicles pulling out without warning

Faults you should avoid

Avoid causing other vehicles to

- Slow down
- Swerve
- Stop

Avoid

- Passing too close to parked vehicles
- Using the size of your vehicle to force other traffic to give way

What the test requires

You should be able to cross the path of oncoming traffic safely and with confidence. You'll need to do this when you

- Turn right at a road junction
- Enter premises on the right-hand side of the road

You should

- Use the MSM/PSL routine
- Position the vehicle as correctly as possible, depending on the size of your vehicle
- Assess accurately the speed of any approaching traffic
- Wait if necessary
- Look into the road entrance that you're about to turn into
- Look out for any pedestrians

How your examiner will test you

For this aspect of driving there isn't a special exercise. Your examiner will watch you and take account of your judgement of oncoming traffic.

Skills you should show

- Make safe and confident decisions about when to turn across the path of vehicles approaching from the opposite direction
- Ensure that the road or entrance is clear for you to enter
- Be confident that your vehicle won't endanger any road user waiting to emerge from the right
- Assess whether it's safe to enter the road entrance
- Show courtesy and consideration to other road users, especially pedestrians

Faults you should avoid

- Cutting the corner
- Overshooting the turn so that the front wheels mount the kerb
- Turning across the path of any oncoming traffic, causing them to
  - slow down
  - swerve
  - stop

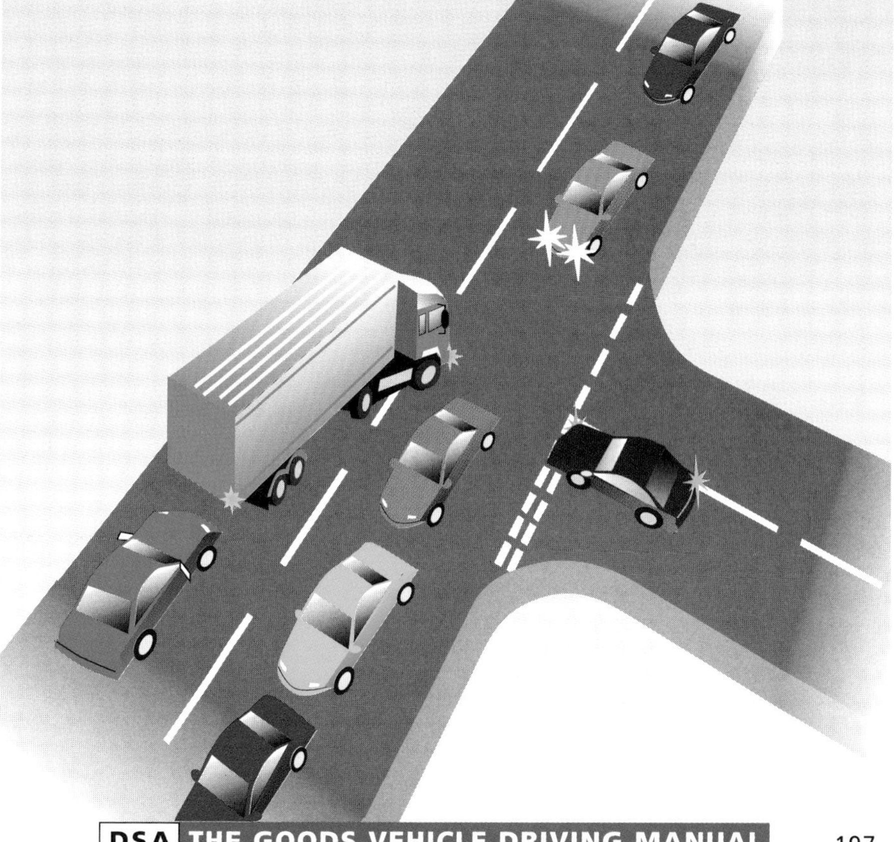

What the test requires

You should

- Recognise the different types of pedestrian crossing
- Show courtesy and consideration towards pedestrians
- Stop safely when necessary

How your examiner will test you

For this aspect of the test there isn't a special exercise. Your examiner will watch you carefully to see that you

- Recognise the pedestrian crossing in good time
- Use the MSM/PSL routine
- Stop when necessary

Skills you should show

Controlled crossings

These crossings may be controlled by traffic signals at junctions or by

- Police officers
- Traffic wardens
- School crossing patrols

Slow down in good time and stop if you're asked to do so.

Zebra crossings

These crossings are recognised by

- Black and white stripes across the road
- Flashing amber beacons on both sides of the road
- Zigzag markings on both sides of the crossing

You should

- Slow down and stop if there's anyone on the crossing
- Slow down and be prepared to stop if there's anyone waiting to cross

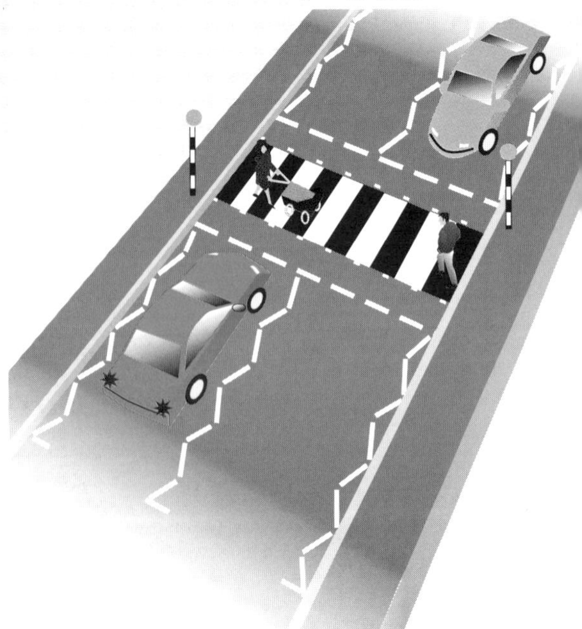

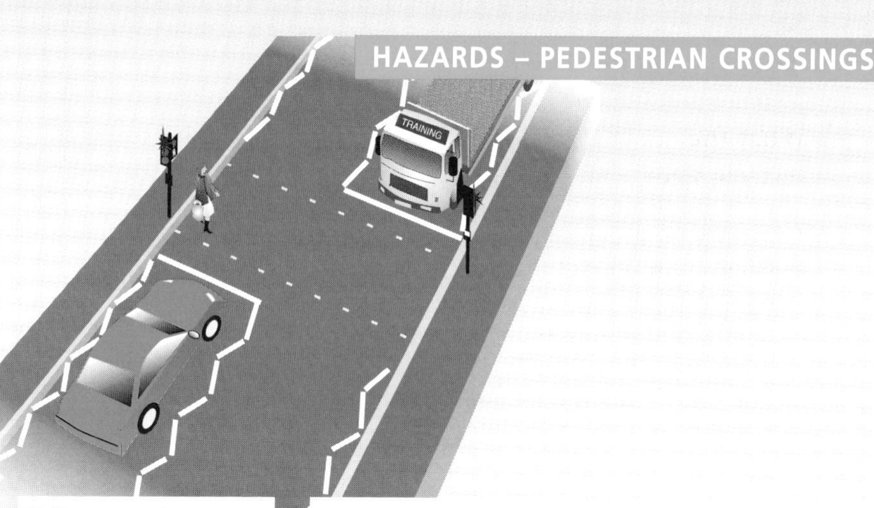

Pelican crossings

These crossings have traffic signals that are activated by pedestrians pressing a button on the panel at either side of the crossing. There's a flashing amber phase that allows pedestrians who are already on the crossing to continue to cross safely.

There are also zigzag lines on each side of the crossing and a stop line at the crossing.

You must

- Stop if the lights are red or steady amber
- Give way to any pedestrians crossing if the amber lights are flashing
- Approach all crossings at a controlled speed
- Stop safely when necessary
- Only move off when it's safe to do so
- Be especially alert
  - near schools
  - at shopping areas
  - when turning at junctions

Puffin crossings

This type of crossing has been installed at a number of selected sights. They have infra-red detectors sited so that the red traffic signal phase may be held until pedestrians have cleared the crossing. No flashing amber is then necessary. The traffic signals operate in the normal sequence.

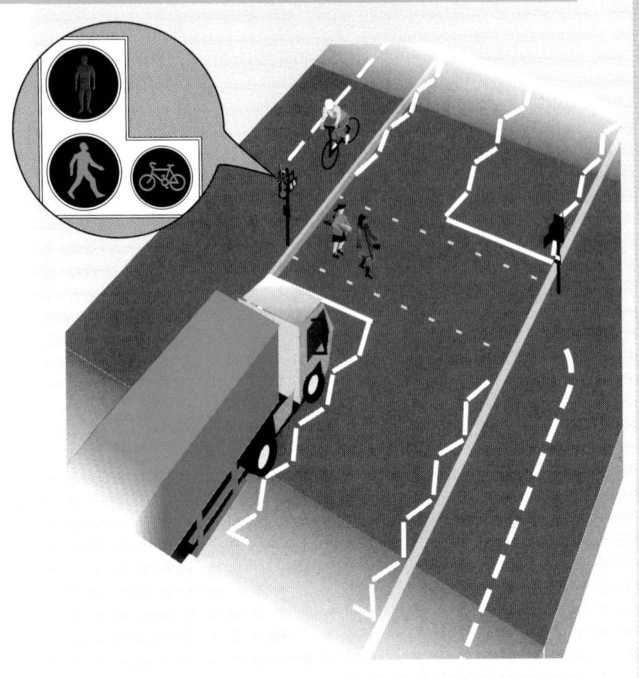

Faults you should avoid

- Approaching any type of crossing at too high a speed
- Driving on without stopping or showing awareness of pedestrians waiting to cross
- Driving onto or blocking a crossing
- Overtaking within the zigzag lines
- Waving pedestrians to cross
- Revving the engine
- Causing unnecessary air brake noise
- Sounding the horn

Toucan crossings

There are a few of these crossings installed throughout the country. They're usually found where there are large numbers of cyclists.

The cyclists share the crossing with pedestrians without having to dismount. They're shown a green cycle light when it's safe to cross.

What the test requires

When you make a normal stop you must be able to select a safe place where you won't

- Cause an obstruction
- Create a hazard
- Be illegally parked

You should stop reasonably close to the kerb.

How your examiner will test you

At some stage during the test your examiner will ask you to pull up on the left at a convenient place.

Skills you should show

When selecting a safe place to stop

- Identify it in good time
- Make proper use of the MSM/PSL routine
- Only stop where you're allowed to do so
- Recognise a place where there aren't parking or stopping restrictions

Faults you should avoid

- Pulling up with late warning to other road users
- Causing danger or inconvenience to other road users
- Not complying with
  - 'no waiting' signs or markings
  - 'no parking' signs or markings
  - 'no stopping' restrictions
- Stopping at or outside
  - school entrances
  - fire or ambulance stations
  - bus stops
  - pedestrian crossings

What the test requires

You should know and be able to demonstrate how to uncouple and recouple your vehicle and trailer safely.

Uncoupling

When uncoupling you should

- Ensure that the brakes are applied on both the vehicle and trailer
- Lower the landing gear and stow the handle away safely
- Turn off any taps fitted to the air lines
- Disconnect the air lines (sometimes referred to as 'Suzies') and stow the lines away safely
- Disconnect the electric line and stow it away safely
- Remove any 'dog clip' securing the kingpin release handle
- Release the fifth wheel coupling locking bar, if fitted
- Drive the tractive unit away slowly, checking the trailer either directly or in the mirrors

If you're driving a rigid vehicle with a trailer you might have to support the trailer drawbar before you pull away.

Recoupling

When recoupling

- Ensure that the trailer brake is applied
- Check that the height of the trailer is correct so that it will receive the unit safely
- Reverse slowly up to the trailer until you hear the kingpin mechanism locking into place
- Select a low gear and try to move forward in order to test that the locking mechanism is secure. Do this twice to make sure
- Ensure that the vehicle parking brake is applied
- Connect any 'dog clip' to secure the kingpin release handle
- Connect the air and electric lines. Turn on taps, if fitted
- Raise the landing gear and stow away the handle
- Release the trailer parking brake
- Start up the engine
- Check that the air is building up in the storage tanks
- Check lights and indicators

How your examiner will test you

You'll normally be asked to uncouple and recouple your vehicle and trailer at the end of the test. Your examiner will ask you to do this where there's safe and level ground.

You'll be asked to

- Demonstrate the uncoupling of your vehicle and trailer
- Pull forward until there's a gap between the vehicle and trailer
- Recouple the vehicle and trailer

Your examiner will expect you to make sure that the

- Coupling is secure
- Lights and indicators are working
- Trailer brake is released

Skills you should show

You should be able to uncouple and recouple your vehicle and trailer

- Safely
- Confidently and in good time
- Showing concern for your own and others' health and safety

Faults you should avoid

When uncoupling

- Uncoupling without applying the brakes on the towing vehicle
- Releasing the trailer coupling without the legs being lowered
- Moving forward before the entire correct procedure has been completed

When recoupling

- Not checking the brakes are applied on the trailer
- Not using good, effective observation around your trailer as you reverse up to it
- Recoupling at speed
- Leaving the cab without applying the vehicle's parking brake

Don't attempt to move away without checking the

- Lights
- Indicators
- Trailer brake function

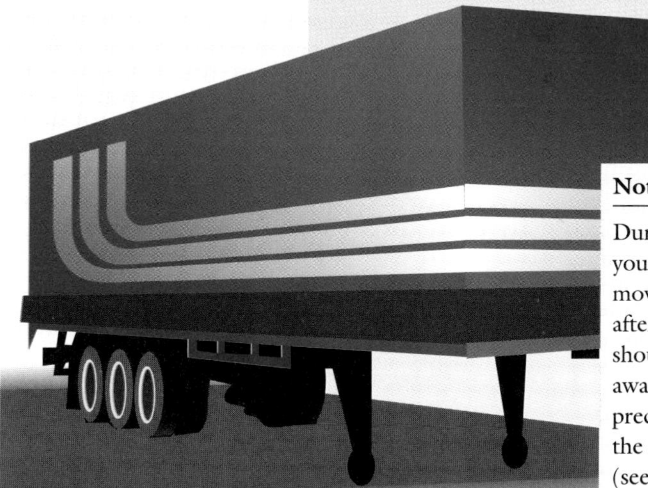

Note:

During the practical test you won't be asked to move your vehicle away after recoupling. You should, however, be aware of the safety precautions given in the official syllabus (see p. 155).

Well done. You've demonstrated that you can drive an unladen LGV to the high standard required to obtain a vocational licence. You should have a sense of pride that should stay with you throughout your professional career.

You'll be given

- A pass certificate (D10V)
- A copy of the driving test report (DLV25), which will show any minor marks that have been marked during the test

You'll also be offered a brief explanation of any minor faults marked. This is to help you to overcome any weaknesses in your driving as you gain experience.

Sign the back of your driving licence and forward it with your pass certificate to

The Vocational Section
DVLA
Swansea
SA99 1BR

as soon as you can.

You should always aim to improve your driving standards. Experience will present situations that perhaps you may not have encountered before. Learn from these and this will increase your ability to become a safe and reliable professional driver. Speak to your trainer about learning to drive laden vehicles.

Your driving hasn't reached the high standard required to obtain the vocational licence. You've made mistakes that either caused or could have caused danger on the road.

Your examiner will give you

- A statement of failure including the driving test report (DLV25A), which will show all the faults marked during the test
- An explanation of why you failed

You should study the driving test report carefully and refer to the relevant sections in this book.

Show the report to your instructor, who will help you to correct the points of failure. Listen to the advice your instructor gives and try to get as much practice as you can before you retake your test.

Right of appeal

Although your examiner's decision can't be altered you have the right to appeal if you consider that your driving test wasn't conducted according to the regulations.

If you live in England or Wales you have six months after the issue of the statement of failure in which to appeal (Magistrates' Courts Act 1952, Ch. 55 part VII, Section 104). If you live in Scotland you have 21 days in which to appeal (Sheriff Court, Scotland Act of Sederunt (Statutory Appeals) 1981).

PART SIX ADDITIONAL INFORMATION

This part offers further information that might be of help for those planning to become professional LGV drivers.

The topics covered

- Disqualified drivers
- DSA services
- DSA Area Offices
- LGV test centres
- Traffic Area Offices
- Other useful addresses
- Minimum test vehicles (MTVs)
- Hazard labels
- Glossary of terms

Retesting once disqualified

Tougher penalties now exist for anyone convicted of certain dangerous driving offences. If a driver is convicted of a dangerous driving offence they'll lose all LGV entitlement.

The decision as to whether they'll be required to undergo an extended car driving test before they're allowed to drive an LGV again rests with the courts.

An LGV driving licence can't stand on its own. You must also possess a valid full driving licence for category B (that is, a car licence). If you lose your category B licence entitlement you also lose your LGV licence.

Applying for a retest

A person subject to a category B retest can apply for a provisional licence at the end of the period of disqualification.

The normal rules for provisional licence-holders apply

- The driver must be supervised by a person who's at least 21 years old and has held (and still holds) a full licence for at least three years for the category of vehicle being driven
- L plates (or D plates, if you wish, in Wales) must be displayed to the front and rear of the vehicle
- Driving on motorways isn't allowed
- LGVs may not be driven on a provisional car licence (category B)

Having passed a category B extended test, an LGV driver has to apply to the Traffic Commissioner for their LGV licence. The return of the licence is at the discretion of the Traffic Commissioner, who can order a retest if he or she considers that it is necessary. A further retest is not ordered in all cases.

Service standards

The Driving Standards Agency (DSA) is committed to providing a high-quality service for all its customers. If you would like information about our standards of service please contact

Customer Services Manager
Driving Standards Agency
Stanley House
Talbot Street
Nottingham
NG1 5GU

Tel: 0115 901 2515/6

Compensation code

The DSA will normally refund the fee, or give a free re-booking, in the following cases

- Where an appointment is cancelled by the DSA – for any reason
- Where an appointment is cancelled by the candidate, who gives at least ten clear working days' notice
- Where the candidate keeps the test appointment, but the test doesn't take place or isn't completed for reasons not attributable to him or her nor to any vehicle provided for the test by the candidate

In addition, the DSA will normally consider reasonable claims from the candidate for financial loss or expenditure unavoidably and directly incurred by him or her as a result of the DSA cancelling the test at short notice (other than for reasons of bad weather). For example, a claim for the commercial hire of the vehicle for the test will normally be considered. Applications should be made to the Area Office where the test was booked.

This compensation code doesn't affect your existing legal rights.

Complaints guide

The DSA aims to give its customers the best possible service. Please tell us

- When we've done well
- When you aren't satisfied

Your comments can help us to improve the service we have to offer.

If you have any questions about how your test was conducted please contact the local Supervising Examiner, whose address is displayed at your local driving test centre. If you're dissatisfied with the reply or you wish to comment on other matters you can write to the Regional Manager. (See the list of Area Offices at the back of this book.)

If your concern relates to an Approved Driving Instructor you should write to

The Registrar of Approved
Driving Instructors
Driving Standards Agency
Stanley House
Talbot Street
Nottingham
NG1 5GU

Finally, you can write to

The Chief Executive
Driving Standards Agency
Stanley House
Talbot Street
Nottingham
NG1 5GU

None of this removes your right to take your complaint to

- Your Member of Parliament, who may decide to raise your case personally with the DSA Chief Executive, the Minister, or the Parliamentary Commissioner for Administration (the Ombudsman)
  WK Reid CB
  Church House
  Great Smith Street
  London SW1P 3BW
  Tel: 0171 276 2003/3000
- A magistrates court (in Scotland to the Sheriff of your area) if you believe that your test wasn't carried out according to the regulations

Before doing this you should seek legal advice.

Driving Standards Agency

DSA Head Office

Stanley House
Talbot Street
Nottingham
NG1 5GU

Tel: 0115 901 2500

National Telephone Numbers

For telephone bookings by Credit or Debit Card and other enquiries: 0870 01 01 372
Fax applications: 0870 01 02 372
Welsh speakers: 0870 01 00 372
Minicom users: 0870 01 07 372

DSA Area Offices

London and the South-East
DSA
PO Box 289
Newcastle-upon-Tyne
NE99 1WE

Midlands and Eastern
DSA
PO Box 287
Newcastle-upon-Tyne
NE99 1WB

Wales and Western
DSA
PO Box 286
Newcastle-upon-Tyne
NE99 1WA

Northern
DSA
PO Box 280
Newcastle-upon-Tyne
NE99 1FP

Scotland
DSA
PO Box 288
Newcastle-upon-Tyne
NE99 1WD

You'll find LGV driving test centres in the following places. Some test centres aren't open full-time. Contact the national telephone numbers for details of the full office address and telephone number.

Scotland

Aberdeen
Bishopbriggs (Glasgow)
Connel
Galashiels
Inverness
Kilmarnock
Kirkwall
Lerwick
Livingstone (Edinburgh)
Locharbriggs (Dumfrieshire)
Machrihanish (Kintyre)
Oban
Perth
Port Ellen
Stornoway
Wick

Northern

Beverley
Bredbury
Carlisle
Darlington
Grimsby
Heywood
Kirkham
Newcastle
Patrick Green
Sheffield
Simonswood
Steeton
Upton
Walton

Wales and Western

Bristol
Caernarfon
Camborne
Exeter
Gloucester
Llantrisant
Neath
Plymouth
Pontypool
Poole
Taunton
Swindon
Withybush
Wrexham

Midlands and Eastern

Birmingham
Chelmsford
Derby
Featherstone
Harlescott
Ipswich
Leicester
Leighton Buzzard
Norwich
Nottingham
Peterborough
Swynnerton
Weedon

London and the South-East

Croydon
Enfield
Guildford
Purfleet
Yeading

Traffic Area Offices

Scotland

Argyll House
3 Lady Lawson Street
Edinburgh EH3 9SE

Tel: 0131 529 8500

Area covered:
All Scotland and the Islands

North-Eastern and North Western

Hillcrest House
386 Harehills Lane
Leeds LS9 6NF

Tel: 0113 283 3533

Area covered:
Cheshire
Cleveland
Cumbria
Durham
Greater Manchester
Humberside
Lancashire
Merseyside
North Yorkshire
Northumberland
Nottinghamshire
South Yorkshire
Tyne and Wear
West Yorkshire

West Midlands

Cumberland House
200 Broad Street
Birmingham B15 1TD

Tel: 0121 608 1000

Area covered:
Hereford
Worcestershire
Shropshire
Staffordshire
Warwickshire
West Midlands

Eastern

Terrington House
13–15 Hills Road
Cambridge CB2 1NP

Tel: 01223 532037

Area covered:
Bedfordshire
Buckinghamshire
Cambridgeshire
Essex
Hertfordshire
Leicestershire
Lincolnshire
Norfolk
Northamptonshire
Suffolk

Wales

Use address as for West Midlands

Area covered:
Clwyd
Dyfed
Gwent
Gwynedd
Mid Glamorgan
Powys
South Glamorgan
West Glamorgan

Western

The Gaunt's House
Denmark Street
Bristol BS1 5DR

Tel: 0117 975 5000

Area covered:
Avon
Berkshire
Cornwall
Devon
Dorset
Gloucestershire
Hampshire
Isle of Wight
Oxfordshire
Somerset
Wiltshire

South-Eastern and Metropolitan London

Ivy House
3 Ivy Terrace
Eastbourne BN21 4QT

Tel: 01323 451400

Area covered:
East Sussex
Greater London
Kent
Surrey
West Sussex

CENTREX

High Ercall
Telford
Shropshire TF6 6RB

Tel: 01952 777777
Fax: 01952 777799

City and Guilds London Institute

1 Giltspur Street
London EC1 9DD

Tel: 0171 294 2468

Department of Transport Mobility Advice and Vehicle Information Service (MAVIS)

DOT
'O' Wing, MacAdam Avenue
Old Wokingham Road
Crowthorne RG45 6XD

Tel: 01344 661000

Driver and Vehicle Licensing Agency (DVLA)

The Vocational Licence Section
Swansea SA99 1BR

Tel: 01792 772151

DVLA Customer Enquiry Unit

Swansea SA6 7JL

Tel: 01792 772151
Minicom: 01792 782756
Fax: 01792 783071
(Service available between 8.15 and 4.30 Monday–Friday)

DVLA Drivers' Medical Branch

Swansea SA99 1TU

Tel: 01792 304000

Freight Transport Association Ltd

Hermes House
St John's Road
Tunbridge Wells
Kent TN4 9UZ

Tel: 01892 526171

Health and Safety Executive Agency - Public Enquiry Point

Broad Lane
Sheffield S3 7HQ

Tel: 0541 545500

(See your telephone book for details of your local HSE office.)

Road Haulage Association Ltd (RHA)

Roadway House
35 Monument Hill
Weybridge
Surrey KT13 8RN

Tel: 01932 841515

Royal Society for the Prevention of Accidents (RoSPA)

Edgbaston Park
353 Bristol Road
Birmingham B5 7ST

Tel: 0121 248 2000

Categories of LGV licences

Category	Description
C1	Medium-sized goods vehicle 3.5–7.5 tonnes
C1 + E	Medium-sized goods vehicle 3.5–7.5 tonnes plus a trailer above 750kg
C	Large goods vehicle above 3.5 tonnes
C + E	Large goods vehicle above 3.5 tonnes with a trailer above 750kg

Minimum test vehicles (MTVs)

Category	Description
C1	Vehicle of at least 4 tonnes and capable of 80 kph (50 mph)
C1 + E	Category C1 test vehicle with trailer of at least 2 tonnes and a combined length of at least 8 metres
C	Vehicle of at least 10 tonnes, at least 7 metres in length and capable of at least 80 kph (50 mph)
C + E	An articulated vehicle of at least 18 tonnes and 12 metres in length, capable of 80 kph (50mph); or a combination of a category C test vehicle with a trailer of at least 4 tonnes and a platform of at least 4 metres in length and with a combined weight of at least 18 tonnes and a combined length of 12 metres. The trailer is to operate with the appropriate service brakes and a heavy duty coupling arrangement suitable for the weight.

All the above weights refer to the maximum authorised mass (MAM).

FLAMMABLE GAS
DANGEROUS WHEN WET
HARMFUL
STOW AWAY FROM FOODSTUFFS
DANGEROUS SUBSTANCE
COMPRESSED GAS
TOXIC GAS
DANGEROUS SUBSTANCE
CORROSIVE
OXIDIZING AGENT

Ready reckoner: metric measurements

This list of metric measurements should prove useful if you want to practise the reversing exercise.

To calculate the reversing area's layout identify the length of your vehicle in the left-hand columns and scan across to the right-hand columns for the relevant cone measurements. The cone positions are relative to the base line Z (see diagram on p. 161).

Metres	Feet	Cone A	Cone B
4.50	14.8	22.5	13.5
4.75	15.6	23.8	14.3
5.00	16.4	25.0	15.0
5.25	17.2	26.3	15.8
5.50	18.0	27.5	16.5
5.75	18.9	28.8	17.3
6.00	19.7	30.0	18.0
6.25	20.5	31.3	18.8
6.50	21.3	32.5	19.5
6.75	22.1	33.8	20.3
7.00	23.0	35.0	21.0
7.25	23.8	36.3	21.8
7.50	24.6	37.5	22.5
7.75	25.4	38.8	23.3
8.00	26.2	40.0	24.0
8.25	27.1	41.3	24.8
8.50	27.9	42.5	25.5
8.75	28.7	43.8	26.3
9.00	29.5	45.0	27.0
9.25	30.3	46.3	27.8
9.50	31.2	47.5	28.5
9.75	32.0	48.8	29.3
10.00	32.8	50.0	30.0
10.25	33.6	51.3	30.8
10.50	34.4	52.5	31.5
10.75	35.3	53.8	32.3
11.00	36.1	55.0	33.0
11.25	36.9	56.3	33.8
11.50	37.7	57.5	34.5
11.75	38.5	58.8	35.3
12.00	39.4	60.0	36.0
12.25	40.2	61.3	36.8
12.50	41.0	62.5	37.5
12.75	41.8	63.8	38.3
13.00	42.7	65.0	39.0
13.25	43.5	66.3	39.8
13.50	44.3	67.5	40.5
13.75	45.1	68.8	41.3
14.00	45.9	70.0	42.0
14.25	46.8	71.3	42.8
14.50	47.6	72.5	43.5
14.75	48.4	73.8	44.3
15.00	49.2	75.0	45.0
15.25	50.0	76.3	45.8
15.50	50.9	77.5	46.5
15.75	51.7	78.8	47.3
16.00	52.5	80.0	48.0
16.25	53.3	81.3	48.8
16.50	54.1	82.5	49.5
16.75	55.0	83.8	50.3
17.00	55.8	85.0	51.0
17.25	56.6	86.3	51.8
17.50	57.4	87.5	52.5
17.75	58.2	88.8	53.3
18.00	59.1	90.0	54.0
18.25	59.9	91.3	54.8

A **ABS** Anti-lock braking system (developed by Bosch) that uses electronic sensors to detect when a wheel is about to lock, releases the brakes sufficiently to allow the wheel to revolve, then repeats the process in a very short space of time – thus avoiding skidding.

ADR Abbreviation used for European rules for the transport of hazardous materials by road.

Air suspension system This uses a compressible material (usually air), contained in chambers located between the axle and the vehicle body, to replace normal steel-leaf spring suspension. Gives an even load height (empty or laden) and added protection to fragile goods in transit.

Axle weights Limits laid down for maximum permitted weights carried by each axle – depending on axle spacings and wheel/tyre arrangement. (Consult regulations, charts or publications that give the legal requirements.)

B **BS 5750** British Standards code relating to quality assurance adopted by vehicle body-builders, recovery firms, etc.

C **C & U (Regs)** Construction and Use regulations that set out in law specifications that govern the design and use of goods vehicles.

CAG Computer-aided gearshift system developed by Scania that employs an electronic control unit combined with electropneumatic actuators and a mechanical gearbox. The clutch is still required to achieve the gear change using an electrical gear lever switch.

City of London Security Regulations Anti-terrorist measures which mean that access to the City of London is restricted to only seven access points, involving closure of several other roads. Full details can be obtained from the Metropolitan Police.

COSHH Regulations 1988 The Control of Substances Hazardous to Health Regulations 1988 place a responsibility on employers to make a proper assessment of the effects of the storage or use of any substances that may represent a risk to their employees' health. (Details can be obtained from the Health and Safety Executive.)

CPC Certificate of Professional Competence indicates that the holder has attained the standards of knowledge required in order to exercise proper control of a transport business (and is required before an operator's licence can be granted).

Printed in The United Kingdom for The Stationery Office J0080081 C250 05/99 63789

Cruise control A facility that allows a vehicle to travel at a set speed without use of the accelerator pedal. However, the driver can immediately return to normal control by pressing the accelerator or brake pedal.

D

Diff-lock A device by which the driver can arrange for the power to be transmitted to both wheels on an axle (normally rotating at different speeds when the vehicle is cornering, for example), which increases traction on surfaces such as mud, snow, etc.

Double de-clutching A driving technique employed that allows the driver to adjust the engine revs to the road speed when changing gear. The clutch pedal is released briefly while the gear lever is in the 'neutral' position. When changing down, engine revs are increased to match the engine speed to the lower gear in order to minimise the work load being placed on the synchro-mesh mechanism.

Drive-by-wire Modern electronic control systems that replace direct technical linkages.

E

Electronic engine management system This system monitors and controls both fuel supply to the engine and the contents of the exhaust gasses produced. The system is an essential part of some speed-retarder systems.

Electronic power shift A semi-automatic transmission system, developed by Mercedes, that requires the clutch to be fully depressed each time a gear change is made. This system then selects the appropriate gear.

G

GCW Gross combination weight, applying to articulated vehicles.

Geartronic A fully automated transmission system developed by Volvo. There's no clutch pedal. Instead, there's an additional pedal operating the exhaust brake.

GTW Gross train weight, applying to drawbar combinations.

GVW Gross vehicle weight, applying to solo rigid vehicles and tractor units.

H

HSE The Health and Safety Executive. The HSE produce literature that provides advice and information on health and safety issues at work.

I

Inter-modal operations Combined road and rail operations for the movement of goods where the 44 tonnes weight limit is authorised – subject to certain conditions.

J

Jake brake A long-established system of speed retarding that alters the valve timing in the engine. In effect, the engine becomes a compressor and holds back the vehicle's speed.

K

Kerb weight The total weight of a vehicle plus fuel, excluding any load (or driver).

L

LA The Licensing Authority, which is an official appointed to act on behalf of the Traffic Commissioners for a Traffic Area.

Laminated A process where plastic film is sandwiched between two layers of glass so that an object upon striking a windscreen, for example, will normally indent the screen without large fragments of glass causing injury to the driver.

Lifting axle An axle that may be lowered or raised, depending on whether the load is required to be distributed to include the additional axle or the vehicle is running unladen. Such axles may be driven, steered or free-running.

LNG Liquefied (compressed) Natural Gas.

Load-sensing valve A valve in an air brake system that can be adjusted to reduce the possibility of wheels locking when the vehicle is unladen.

London lorry ban Night-time and weekend ban on lorries over 16.5 tonnes maximum gross weight, applying to most roads in Greater London other than trunk roads and exempted roads. All vehicles other than special types or those concerned with safety or emergency operations must display a permit and exemption plate if they're to be used in the restricted areas.
Note: Not all London boroughs still operate the scheme.

LPG Liquefied (compressed) Petroleum Gas.

M

MAM Maximum authorised mass, also known as maximum authorised weight or gross vehicle weight.

P

Plated Department of Transport regulations for recording and displaying information relating to dimensions and weights of goods vehicles, indicating maximum gross weight, maximum axle weight and maximum train weight. In the case of trailers the plate indicates maximum gross weight and maximum axle weight for each axle. (This is in addition to any manufacturer's plate that's fixed to the vehicle or trailer.)

R **Range change** Gearbox arrangement that permits the driver to select a series of either high or low ratio gears depending on the load, speed and any gradient being negotiated. Effectively doubles the number of gears available (frequently up to a total of 16 gears, including crawler gears).

Red Routes Approximately 300 route miles in the London area are to become subject to stringent regulations restricting stopping, unloading and loading.

Re-grooving A process permitted for use on tyres for vehicles over an unladen weight of 2,540kg, allowing a new tread pattern to be cut into the existing tyre surface (subject to certain conditions).

Retarder An additional braking system that may be either mechanical or electrical. Mechanical devices either alter the engine exhaust gas flow or amend the valve timing (creating a 'compressor' effect). Electrical devices comprise an electromagnetic field energised around the transmission drive shaft (more frequently used on passenger vehicles).

S **SAMT** Semi-automatic transmission system in which the clutch is only used when starting off or stopping.

Selective or block change A sequence of gear-changing omitting intermediate gears. Sometimes known as selective gear-changing.

Splitter box Another name for a gearbox with high and low ratios that effectively doubles the number of gears available.

T **Tachograph** A recorder indicating vehicle speeds, duration of journey, rest stops, etc. Required to be fitted to specified vehicles.

TBV Initials of the French (Renault) semi-automatic transmission system that employs a selector lever plus visual display information.

Thinking gearbox The term used to describe a fully automated gearbox that selects the appropriate gear for the load, gradient and speed, etc. by means of electronic sensors.

Toughened safety glass The glass undergoes a heat treatment process during manufacture so that in the event of an impact (a stone, etc.) on the windscreen it breaks up into small blunt fragments, thus reducing the risk of injury. An area on the windscreen in front of the driver is designed to give a zone of vision in the event of such an impact.

Trailer swing This occurs when severe braking causes partial loss of control as the rear wheels of a semi-trailer lock up on an articulated vehicle.

Turbo-charged Forced air (fan-driven) mixed with fuel to give increased engine performance.

Turbo-cooled Forced air (fan-driven) in addition to a liquid engine coolant system.

Two-speed axle An electrical switch actuates a mechanism in the rear axle that doubles the number of ratios available to the driver.

U **Unloader valve** A device fitted to air brake systems, between the compressor and the storage reservoir, pre-set to operate as sufficient pressure is achieved and allowing the excess to be released (often heard at regular intervals when the engine is running).

V **VEL** Vehicle Excise Licence or road fund licence.

VRO Vehicle Registration Office, dealing with matters relating to registration of goods vehicles, taxation and licensing.

ABS (anti-lock braking systems) 46-7
  checking 47
Acceleration, sudden 19, 20, 24
Accidents 14, 26, 112, 124-33, 139
  at the scene of 125
  reporting 133
Age, minimum 6
Air brake systems 19, 50, 147, 167
Air lines ('Suzie lines') 48, 50, 202
Air reservoirs 50
Air suspension 31, 37
Alcohol 71
Animals 82, 184
Anti-lock braking systems (ABS) 46-7
  checking 47
Anti-roll stability 31
Anti-theft measures 79
Anticipation 81, 105, 178
Articulated vehicles
  car transporters 33
  characteristics 30
  jack-knifing 21
  part-loads 38
  provisional licence 7
  stability 30-1, 33, 34, 38
  tankers 30-2
  weight limits 38
Asbestos 78
Asleep at the wheel, falling 70, 93
Attitude 2, 14-18
Audible warning systems 61
Automatic transmission, licence restrictions for 7
Auxiliary lighting 95
Awareness 81, 84, 87, 105, 178
Axle weight limits 38

Battens, load securing 54
Behaviour, appropriate 14-16, 151
Bends, speed on 25
Bleepers, reversing 61
Blind spots 18, 86
Blow-outs 134
Bodies
  demountable 34
  double-decked 34
Boredom, avoiding 70
Box vans 28, 29
Brake (*see also* Brakes, Braking)
  fade 24
  hand 168
  lines, connecting 48, 49, 51
  parking 46, 49
  secondary 46
  service 46, 49
Brakes 46-51 (*see also* Brake, Braking)
  applying 46
  overheating 24
  poorly adjusted 21
  pumping 19
  use of, for test 167-8
Braking (*see also* Brake, Brakes)
  cadence 46
  distance 24
  effort 24
  exercise for test 137, 157, 163, 167
  harsh 19, 20, 21, 24
  in good time 19, 27
  loss of control 25
  sudden 19, 20
Braking systems 46-51, 139 (*see also* Brake, Brakes)
  air 19, 50, 147, 167
  anti-lock (ABS) 46-7
    checking 47
  connecting and disconnecting 48
  controls 50
  endurance ('retarders') 21, 46, 47-8, 119, 151

hydraulic 19
inspection 50
maintenance 50, 116
safety 49, 50
taps 50
three-line 48, 51
two-line 48, 49
Breakdowns 134
at night 99
on motorways 112-13
Bridges, collision with 41
Buffeting 17, 18
Building sites 118
Built-up roads, care required 17
at night 96

Cadence braking 46
Cameras 75
Car transporters, articulated 33
Caravans, passing 17
Carriageway markings, reflective 120
Cars, passing 17
Cats' eyes 120
Centre of gravity 22, 30
Centrifugal force 25
Chains, load securing 54
Children 82
carrying 149
Chocks, load securing 54
Clearances (*see* Limits)
Compressed gas loads 32
Construction sites 118
Contact lenses 9
Containers, securing 56
Contraflows 111
Control
loss of 19, 25
maintaining 27, 81
Controls, understanding, for test 150-51, 164-170
Coolant 102
Courtesy 16
Crosswinds 29, 56, 108 (*see also* Winds, high)
Curtain sides 56
Cyclists 17, 18, 30, 82, 84, 87, 124, 184, 193

D plates 6, 149
D1 form 6, 9
D4 form 10
Dangerous goods 60
Deceleration, sudden 20
Demountable bodies 34
Diesel fuel, 'red' 72
Diesel spillages 59
Diff-lock 118-19
Disability 8, 11, 138
Disqualified drivers 207
Distances, safe 88
DL26 test application form 140
'Dolly' knots 54, 55
Double-decked bodies 34
Drinking and driving 71
Driver Training Regulations (DTR) 60
Drivers' hours of work 62-9, 139
catching up on reduced rest 67
daily driving 66
daily duty 68
definitions 66, 69
domestic 68-9
driving limits 68
exemptions 62-4, 68
mixed EC and domestic driving 69
records 68-9
rest periods 66, 67, 100
two or more drivers 67
weekly driving 66
Drivers Medical Unit 8, 9

Driving
at night 93-9
built-up areas 96
rural areas 97
different sized vehicles 19
forces (*see* Forces)
regulations 74-7
skills when coming across 12, 81-3
animals 82
children 82
cyclists 82
elderly people 83
horses 82
learner drivers 83
other road users 82-3
test (*see* Test)
Driving Standards Agency (*see* DSA)
Drugs 71
DSA (Driving Standards Agency) v, 208-10
DVLA (Driver and Vehicle Licensing Agency) 6

EC drivers' hours rules 62-4, 66-7, 69
EC legislation re. speed limiters 73
Elderly people 83
Electric shock 132
Electrical systems 32
Electricity cables over roads 40
End-tipper vehicles, 23, 58
Endurance braking systems ('retarders') 46, 47-8
Energy, kinetic 24
Environmental impact 57-61
Eyesight test 9

Falling asleep at the wheel 70, 93
Fire 32, 127-8
extinguishers 128
First Aid 130-1
bleeding 131
breathing stopped 130-1
shock 131
electric 132
Flashing headlights 16, 92
Fog 105, 110, 120-1
at night 97
lights, high-intensity rear 105, 110, 117, 120
Force, centrifugal 25
Force of gravity 22, 30
Forces
acting on a load 24, 25
acting on a tanker 30
affecting vehicle 22-5
driving 19-27
kinetic energy 24
momentum 24
Form
D1: 6, 9
D4: 10
DL26: 140
Friction 20
Fuel 102, 116
Fuel consumption 58
Fuels, alternative 59

Gas loads 32
Gear-changing 21, 171
Glasses 9
Gradients, effects of gravity on 22
Gravity 22-3, 30

Handbrake 168
Handling characteristics 26, 28-35, 152 (*see also* Vehicle characteristics)
Handling, erratic 20
Hazard labels 216
Hazardous goods 32, 60, 99

accidents involving 126, 127
Hazards 32, 182-200
animals 82, 184
at night 96-7
correct routine 182
crossing other traffic 196-7
cyclists 184, 193
horses 184, 193
junctions 186-93
lane discipline 185
meeting other vehicles 195
motorcyclists 184
other road users 184
overtaking 194
passing other vehicles 195
pedestrian crossings 198-200
positioning 185
road surfaces 193
roundabouts 188-93
(*see also* Roundabouts)
mini- 192-3
multiple 192
skills required 181, 182, 183, 184
turns, left or right 189
Headlights 16
flashing 16, 92
Health 8, 10, 11, 70, 100, 150
Health and Safety 78
at Work Act 55, 60
Height
guide 42
limits 39-42
notice 39
of vehicle 33
warning devices 39
Highway Code 12, 16, 92, 136, 149, 176
Horn 16, 61, 92
Horses 17, 82, 184, 193
Hours of work (*see* Drivers' hours of work)
Hydraulic braking systems 19

Icy weather 109, 117
Illness 8, 10, 11, 70, 100, 150
Inflammable loads 32
Instructors 2
Instruments 102
Intimidation 14-15, 16

Jack-knifing 21
Junctions 87, 186-93

Kerb, running over 22, 30, 85
Kinetic energy 24
Knots, 'dolly' 54, 55

L plates 6, 149
Lane discipline 104, 185
Learner drivers 83
Legal requirements 62-9, 149-50
Length limits 45
LGV driving test (*see* Test)
LGVs
driving large and small, differences 19
understanding 19
Licence, applying for 6
Licences
categories of LGV 215
types of provisional 6-7
Lighting, vehicle 95, 97, 98, 102
Lighting-up times 94
Lights 102
in fog 120
traffic 90-1, 177
Limits
height 39-42
knowledge required 37

Limits (*cont'd*)
- length 45
- vehicle 36-45
- weight 37, 38
- width 44

Load
- distribution 38
- forces acting on a 24, 25
- movement 20
- restraint 52-6
- securing a 52-6, 139
- types of 53

Loads
- compressed gases 32
- fluid 32
- fragile 57
- heavy objects 53, 54
- high 42
- inflammable 32
- logs 54
- meat carcasses 34
- multi-drop 56
- over-height 40
- part- 38
- plastic sacks 53
- shedding 26
- tree trunks 54
- tubular 53
- vehicles 53
- wide 44

Long wheel-base rigid vehicles 28
Loss of control 19, 25

Maintenance 58, 101-2
Making progress 104, 179
MAM (Maximum Authorised Mass) 37
Maximum Gross Weight (MGW) 37
Medical
- examination 10
- requirements 8
- standards 11

Medicines 71
MGW (Maximum Gross Weight) 37
Minimum test vehicles (MTVs) 145, 215
Mirrors 18, 84, 85, 86, 101, 175
- nearside 17, 84, 85, 123

Momentum 24
Motorcyclists 17, 87, 184
Motorway driving, preparing for, 100-2
Motorways, driving on
- accidents 112, 125
- awareness (anticipation) 105
- breakdowns 112-13
- contraflows 111
- crawler lanes 104
- crosswinds 108
- diversions 106
- end of 115
- fog 105, 110
- fog lights, high-intensity rear 105, 110
- ice and frost 109
- joining 103
- lane discipline 104
- leaving 114-15
- queues 115
- rain 108
- roadworks 111
- separation distance 105, 108
- signs and signals 106, 112
- slip roads 63, 104
- speed, reducing when leaving 115
- telephones 113
- weather conditions 108-10

Moving off 170-2
MTVs (minimum test vehicles) 145, 215

Night
  driving 93-9
  hazards 96-7
  vision 94, 96

Observation, effective 84, 88
Observation at junctions 87
Oil 102, 116
Overhead clearances 40
Over-height loads 40
Overloading 38
Overtaking 84, 139, 194
  at night 98

Parking 201
  at night 94, 96
  brake 46, 49
Part-loads 38
Pedestrian crossings 198-200
Pedestrians 17, 30, 82, 87, 124
Pelican crossings 199
Penalty points 75
Pendulum effect 34
Planning ahead 27, 58, 81, 122
  articulated vehicles 30
Police, notifying 41, 44, 99, 133
Positioning 185
Power-assisted steering 168
Professional standards 12-13
Provisional licence 6-7
Puffin crossings 199

Rail bridges, collision with 41
Rain, heavy 108, 117, 118
Recovery agencies 99
Recovery vehicles 38
'Red' diesel fuel 72
Red Routes 76-7
  signs 77, 154
Reflective studs and markings 120
Refrigerated vehicles 34
Regulations
  Health and Safety 78
  vehicle 72-3
Regulations, driving 70-1, 74-7
  knowledge required 37
Reporting accidents 133
Rest periods 66, 67, 100
Restrictions (*see* Limits)
Retaliation 15
'Retarders' 21, 46, 47-8, 119, 151
Retesting once disqualified 207
Reversing, audible warning systems 61
Reversing, exercise for test 137, 161-2
Road
  conditions 20, 153
  friendly suspension 37, 57
  procedure for test 154-8
  surface 20
  user behaviour 14-16, 151
Roll-over 30-1
Ropes, load securing 54
Roundabouts 188-93 (*see also* Hazards)
  going ahead 190
  turning full circle 191
  turning left 189
  turning right 191
Route planning 42

Safe distances 88
Safe working practice 78
Safety 55, 60, 78
  checks 150-51
Secondary brake 46
Security alarm system 61
Semi-trailers, specialised 57 (*see also* Trailers)

Separation distance 88, 89, 117, 118, 121, 181
at night 98
on motorways 105, 108
Service brake 46, 49
Sheet securing 55
Sheeting to cover a load 55, 58
Shopping areas, driving in 17
Short wheel-base vehicles 28
Side marker boards 44
Side-tipper vehicles 23, 58
Signalling 92, 176-7
flashing headlights 16, 92
Signals and signs 106, 112, 176-7
Skidding 20
Snow 118-19
Speed 15, 25, 27, 180
limiters 73, 139, 151
Speeding offences 75
Spray-suppression equipment 102, 108
Stability 19, 22, 31
of articulated vehicles 30-1, 33, 34, 38
of car transporters 33
of tipper vehicles 23
Standards, professional 12-13
Steering
power-assisted 168
sudden movement of 19, 20, 21
use of, for test 168
Stopping at a safe place 156, 201
Straps, load securing 54
Supervision 6
Suspension 31
road-friendly 37, 57
'Suzie' lines (air lines) 48, 50, 202
'Swan-neck' turn 30

Tachograph
calibration 64
chart inspections 65
charts 64
EC rules, exemptions from 62-4
recording information 65
symbols 65
Tachographs 62, 64-5
Tailgating 14
Tankers, articulated 30-2
Tarpaulin sheeting 55
Test, LGV driving
anticipation (awareness) 178
applying 140-1
appointments 141, 208
awareness (anticipation) 178
before 147
booking 140-1
braking exercise 137, 157, 163, 167
categories 6, 7, 145, 149, 215
centres 211
controls, knowledge of 150-51, 164-70
documents 142
examiners 136-7
failing 205
gear-changing 171
handbrake 168
hazards (*see* Hazards)
knowledge required 149
language difficulties 138
legal requirements 146, 149-50
making progress 179
mirrors 175
moving off 172-4
passing 204
postponing 141
Practical 137-40
preparing for 137-9

ready for 140
recoupling 7, 137, 157-8, 202-3
reversing exercise 137, 161-2
right of appeal 205
road conditions 153
road procedure 154-8
road user behaviour 151
safety checks 150-1, 170
separation distance 181
signals and signs 176-7
rules and regulations 154
special circumstances 138
speed 180
steering 168
stopping at a safe place 201
syllabus 137, 148-58
Theory 141
topics covered 160
traffic signs, rules and regulations 154
trainer booking 140
uncoupling 7, 137, 157, 202-3
vehicle characteristics 152
vehicle control 154-8
vehicle requirements 142-6
vehicles, minimum (MTVs) 145, 215
weather conditions 153
Theft 79
Tipper vehicles 23, 58
Tiredness 70, 93, 100
Toucan crossings 200
Traction 20, 25
Traffic
Area Offices 212-13
rules and signs 154, 177
signals 90-1, 177
Trailer brakes
parking 50
three-line 48, 51
two-line 48, 49
Trailer pivoting points 21
Trailer swing 21
Trailers
dual-purpose 53
jack-knifing 21
specialised semi- 57
uncoupling and recoupling 7, 137, 157-8, 202-3
Transporters, car 33
Turbulence, effects of 17, 18
Turning 87, 186-93
an articulated vehicle 30, 139
loss of control when 25
Two-second rule 89
(*see also* Separation distance)
Tyre checks 101, 116
Tyre failures 134
Tyres, loss of grip 20

Vans, box 28, 29
Vehicle
characteristics 26, 28-35, 152
articulated 30 (*see also* Articulated vehicles)
box vans 28, 29
car transporters, articulated 33
demountable bodies 34
double-decked bodies 34
end-tipper 23, 58
long wheel-base rigid 28
recovery 38
refrigerated 34
short wheel-base 28
side-tipper 23, 58
tankers, articulated 30-2
tipper 23, 58
checks 101-2
for bad weather 116
control 154-8

Vehicle (*cont'd*)
knowledge required for test 150-51
lighting 95, 97, 98, 102
limits 36-45
maintenance 58
regulations 72
Vision
at night 94, 96
zones of 87

Weather conditions 108-10, 153 (*see also* Fog, Rain, Snow, Winds)
bad 116-23
icy 109, 117
vehicle checks 116
Weight
checks 37
limits 37, 38
transference 20
Wheel-lift 30
Wheels, lost 134
Wide loads 44
Width limits 44
Winds, high 56, 122-3, 139 (*see also* Crosswinds)
Windscreen 101, 105, 117

Zebra crossings 198